Dear Mary,

With gratitude for your help in getting my start in science, and best wishes for the new year,

Sincerely yours,

Kim [signature] 1/98

Cell Cycle Regulators in Cancer

Authors *Kiran Musunuru,* Boston, Mass.
Philip W. Hinds, Boston, Mass..

31 figures and 2 tables, 1997

KARGER LANDES SYSTEMS

Basel · Freiburg · Paris · London · New York ·
New Delhi · Bangkok · Singapore · Tokyo · Sydney

Kiran Musunuru, M.D.
Department of Pathology
Harvard Medical School
Boston, Massachusetts, USA

Philip W. Hinds
Department of Pathology
Harvard Medical School
Boston, Massachusetts, USA

Library of Congress Cataloging-in-Publication Data
Musunuru, Kiran.
Cell cycle regulators in cancer / Kiran Musunuru and Philip W. Hinds
Includes bibliographical references and index.
1. Cyclins – Pathophysiology. 2. Cyclin-dependent kinases – Pathophysiology.
3. Carcinogenesis – Molecular aspects. I. Hinds, Philip W. II. Title.
[DNLM: 1. Cyclin-Dependent Kinases. 2. Cyclins. 3. Neoplasms – pathology.
QU 141 M991c 1997]
RC268.5.M87 1997 616.99'407 – dc21
ISBN 3–8055–6561–5 (hard cover)

Drug Dosage. The authors and the publisher have exerted every effort to ensure that drug selection and dosage set forth in this text are in accord with current recommendations and practice at the time of publication. However, in view of ongoing research, changes in government regulations, and the constant flow of information relating to drug therapy and drug reactions, the reader is urged to check the package insert for each drug for any change in indications and dosage and for added warnings and precautions. This is particularly important when the recommended agent is a new and/or infrequently employed drug.

Printed in Switzerland on acid-free paper by Reinhardt Druck, Basel
ISBN 3–8055–6561–5

Contents

Acknowledgments

K.M.: I am deeply indebted to Thom Morris, Robert Farese, Mary Standaert, Jeff Ferrell, and especially my family and those who have encouraged me all along the way.

P.W.H.: I am indebted to Arnie Levine, Bob Weinberg and Peter Howley for their constant advice and encouragement, and to the members of my own lab for their constructive criticism of this work. I am also most grateful for the boundless patience and support of my wife and children.

Chapter I

Introduction

Cancer is a devastating disease, the more so because it strikes unexpectedly, indiscriminately, and usually without any obvious cause. One in five people eventually succumbs to cancer. The high incidence suggests that the development of cancer is closely linked to the body's natural processes. The malignant growth that is associated with the end state of cancer is clearly a perversion of the same replicative mechanisms that cause the trillions of cells of an organism to arise from a single zygote.

A terse yet accurate description of cancer, then, is *improper* cellular proliferation. To understand cancer we must first understand *proper* cellular proliferation. While a basic comprehension of the cell cycle has existed for several decades, the last five years have witnessed the characterization of a host of cell cycle regulators and the clarification of the biochemical processes in which they participate. We are now acquainted with some of the players that control the transitions between various phases of the cell cycle, the precise timing of which are crucial for proper tissue integrity. If in a particular cell one transition after another is passed improperly, without constraint, the cell cycle is driven into continuous "play" and a mass of cells—a tumor—will appear where there should only be a single cell. There is a basic, albeit not simple, relationship between the cell cycle and cancer.

Our desire to understand this relationship is the reason for our interest in those cell cycle regulators involved in cancer. The regulators of primary interest are families of proteins called the cyclins, the cyclin–dependent kinases (CDKs), and the CDK inhibitors (CKIs). Members of the cyclin family are notable for their periodic behavior. They are synthesized more or less continuously such that they accumulate as the cell cycle advances, but at defined points in the cycle (different points for specific cyclins), they undergo rapid, complete destruction. As synthesis continues cyclin levels gradually build up again, marking the passing of the next cell cycle. In some sense the cyclins may be perceived as the internal timekeepers of the cell.

The members of the CDK family, in contrast to cyclins, are generally maintained at constant levels throughout the cell cycle. As protein kinases, however, their activity is variable, for they must be bound in complexes with appropriate cyclin partners to be active. Only during the window of opportunity when the cyclin level has peaked can the CDK, as a component of a cyclin–CDK complex, phosphorylate downstream targets. The subsequent destruction of the cyclin inactivates the CDK. CDKs are also susceptible to post–translational modification by a group of kinases and phosphatases, alternatively enabling or disabling CDK activity in response to signal transduction pathways. Between their dependence on cyclins and their subjection to cellular signals, CDKs are constrained to act during targeted intervals of the cell cycle.

The CKI family, the most recently discovered regulators, constitute yet another control on CDK activity. CKIs inhibit cyclin–CDK complexes by means of direct physical interaction and prevent them from acting upon downstream targets.

To date at least twelve mammalian cyclins, eight cyclin–dependent kinases and seven cyclin–dependent kinase inhibitors have been identified (although the rapid advancement of the field promises that these numbers will quickly increase). The available evidence tells us that the interplay of the cyclins, CDKs and CKIs (as well as protein families that potentially remain to be discovered) is the underpinning of an intricate system of checks and balances that governs the progression of the cell cycle. It should not be surprising, then, that gene mutations in some of these regulators have profound implications of tumorigenicity.

This book is an attempt to provide a succinct description of the connections between cell cycle regulators and oncogenesis as they are presently understood. Chapter 2 sets the stage with a historical perspective of cell cycle research in the context of cyclins and CDKs. Chapter 3 describes the structural similarities and functional paradigms of the cyclin and CDK families. Chapters 4 and 5 turn our attention to the regulation of cyclin–CDK complexes, the first focusing on post–translational modification, the second discussing CKI interactions as well as checkpoint controls. Chapters 6 and 7 situate cyclin–CDK complexes at the center of cell cycle control, examining their relationships with various upstream regulators and downstream targets including signal transducers, tumor suppressor proteins and transcription factors. Chapter 8 summarizes the known correlations between the improper function of cell cycle regulators and the incidence of various types of cancer. Chapter 9 brings our examination to a close by integrating the cyclins, CDKs, CKIs and their associated proteins into an overarching model of cell cycle regulation. We will highlight the vulnerabilities in the cell cycle machinery that make it possible for uncontrolled proliferation to take place, and we will explore possible means of clinical intervention that might someday be developed into potent cancer therapies.

It is impossible to cover in this book every interesting finding relevant to cellular proliferation and oncogenesis. Indeed, a comprehensive review of this rapidly burgeoning field or even subsets of the field could easily fill several books, with at least another book being added to the set every year. We have therefore attempted to establish a framework within which the reader is encouraged to flesh out the details by referring to the excellent reviews listed at the end of this chapter. This framework is intended to convey a conceptual understanding of cell cycle regulation so that we may extend the concepts and study the philosophical implications of these concepts with regards to cancer. Paramount to this approach is an emphasis on the relationships between the cell cycle regulators rather than the individual regulators themselves. Since much of the science that led to the information presented in this book was relational in nature (e.g., protein A was discovered via its *functional interaction* with protein B), we have devoted what may seem like a disproportionate amount of space to chronologically documenting the discovery of the cell cycle regulators. This was in the hope that we would more readily communicate to the reader the important conceptual relationships by showing exactly how researchers proceeded in uncovering them.

Our strategy admittedly makes our investigation susceptible to our biases as to what is more relevant and what is less relevant, but we feel that it is the most reasonable approach in the interest of concision. We offer our humblest apologies to our colleagues whose work we have neglected to include, either by constraints or by oversight. We direct the reader to the following selections of review articles, which together present a far more complete picture than we can achieve in this short space.

Cell Cycle and Cancer

1 Arnold A. The cyclin D1/PRAD1 oncogene in human neoplasia. J Investig Med 1995; 43:543–549.

2 Bates S, Peters G. Cyclin D1 as a cellular proto–oncogene. Semin Cancer Biol 1995; 6:73–82.

3 Biggs JR, Kraft AS. Inhibitors of cyclin–dependent kinase and cancer. J Mol Med 1995; 73: 509–514.

4 Brechot C. Oncogenic activation of cyclin A. Curr Opin Genet Dev 1993; 3:11–18.

5 Cordon–Cardo C. Mutations of cell cycle regulators. Biological and clinical implications for human neoplasia. Am J Pathol 1995; 147:545–560.

6 Della Ragione F, Mercurio C, Iolascon A. Cell cycle regulation and human leukemias: the role of $p16^{INK4}$ gene inactivation in the development of human acute lymphoblastic leukemia. Haematologica 1995; 80:557–568.

7 Devilee P, Schuuring E, van de Vijver MJ et al. Recent developments in the molecular genetic understanding of breast cancer. Crit Rev Oncog 1994; 5:247–270.
8 Dickson C, Fantl V, Gillett C et al. Amplification of chromosome band 11q13 and a role for cyclin D1 in human breast cancer. Cancer Lett 1995; 90: 43–50.
9 Harris AW, Bodrug SE, Warner BJ et al. Cyclin D1 as the putative bcl-1 oncogene. Curr Top Microbiol Immunol 1995; 194:347–353.
10 Hartwell LH, Kastan MB. Cell cycle control and cancer. Science 1994; 266:1821–1828.
11 Hirama T, Koeffler HP. Role of the cyclin–dependent kinase inhibitors in the development of cancer. Blood 1995; 86:841–854.
12 Hunter T, Pines J. Cyclins and cancer. II: Cyclin D and CDK inhibitors come of age. Cell 1994; 79:573–582.
13 Hunter T, Pines J. Cyclins and cancer. Cell 1991; 66:1071–1074.
14 Kamb A. Cell–cycle regulators and cancer. Trends Genet 1995; 11:136–140.
15 Morgan DO. Cell cycle control in normal and neoplastic cells. Curr Opin Genet Dev 1992; 2:33–37.
16 Motokura T, Arnold A. Cyclin D and oncogenesis. Curr Opin Genet Dev 1993; 3:5–10.
17 Okayama H, Nagata A, Igarashi M et al. Mammalian G2 regulatory genes and their possible involvement in genetic instability in cancer cells. Princess Takamatsu Symp 1991; 22:231–238.
18 Pagano M, Draetta G. Cyclin A, cell cycle control and oncogenesis. Prog Growth Factor Res 1991; 3:267–277.
19 Peters G, Fantl V, Smith R et al. Chromosome 11q13 markers and D-type cyclins in breast cancer. Breast Cancer Res Treat 1995; 33:125–135.
20 Peters G. The D-type cyclins and their role in tumorigenesis. J Cell Sci Suppl 1994; 18:89–96.
21 Pines J. Cyclins, CDKs and cancer. Semin Cancer Biol 1995; 6:63–72.
22 Raffeld M, Sander CA, Yano T et al. Mantle cell lymphoma: an update. Leuk Lymphoma 1992; 8:161–166.
23 Schmidt EV. Happenstance, circumstance or enemy action: cyclin D1 in breast, eye and brain. Bioessays 1996; 18:6–8.
24 Schuuring E. The involvement of the chromosome 11q13 region in human malignancies: cyclin D1 and EMS1 are two new candidate oncogenes—a review. Gene 1995; 159:83–96.
25 Sheaff RJ, Roberts JM. Tumor suppression. Lessons in p16 from phylum Falconium. Curr Biol 1995; 5:28–31.
26 Sutherland RL, Hamilton JA, Sweeney KJ et al. Expression and regulation of cyclin genes in breast cancer. Acta Oncol 1995; 34:651–656.
27 Sutherland RL, Watts CK, Musgrove EA. Cyclin gene expression and growth control in normal and neoplastic human breast epithelium. J Steroid Biochem Mol Biol 1993; 47:99–106.
28 Tahara E. Genetic alterations in human gastrointestinal cancers. The application to molecular diagnosis. Cancer 1995; 75:1410–1417.
29 Tripathy D, Benz CC. Activated oncogenes and putative tumor suppressor genes involved in human breast cancers. Cancer Treat Res 1992; 63:15–60.
30 Westermark B, Nister M. Molecular genetics of human glioma. Curr Opin Oncol 1995; 7:220–225.
31 Weinberg RA. How cancer arises. Sci Am 1996; 275:62–70.

Cell Cycle Machinery

1 Charollais RH, Tiwari S, Thomas NS. Into and out of G1: the control of cell proliferation. Biochimie 1994; 76:887–894.
2 Clarke PR. Cyclin–dependent kinases. CAK–handed kinase activation. Curr Biol 1995; 5:40–42.
3 Coleman TR, Dunphy WG. Cdc2 regulatory factors. Curr Opin Cell Biol 1994; 6:877–882.
4 Deshaies RJ. The self–destructive personality of a cell cycle in transition. Curr Opin Cell Biol 1995; 7:781–789.

5 Doree M, Galas S. The cyclin–dependent protein kinases and the control of cell division. FASEB J 1994; 8:1114–1121.
6 Draetta GF. Mammalian G1 cyclins. Curr Opin Cell Biol 1994; 6:842–846.
7 Edgar BA. Cell cycle. Cell–cycle control in a developmental context. Curr Biol 1994; 4:522–524.
8 Elledge SJ, Harper JW. Cdk inhibitors: on the threshold of checkpoints and development. Curr Opin Cell Biol 1994; 6:847–852.
9 Grana X, Reddy EP. Cell cycle control in mammalian cells: role of cyclins, cyclin dependent kinases (CDKs), growth suppressor genes and cyclin–dependent kinase inhibitors (CKIs). Oncogene 1995: 11:211–219.
10 Heichman KA, Roberts JM. Rules to replicate by. Cell 1994; 79:557–562.
11 Hoffmann I, Karsenti E. The role of cdc25 in checkpoints and feedback controls in the eukaryotic cell cycle. J Cell Sci Suppl 1994; 18:75–79.
12 Hunt T. Cyclins and their partners: from a simple idea to complicated reality. Semin Cell Biol 1991; 2:213–222.
13 Kaufmann WK, Paules RS. DNA damage and cell cycle checkpoints. FASEB J 1996; 10:238–247.
14 King RW, Jackson PK, Kirschner MW. Mitosis in transition. Cell 1994; 79:563–571.
15 Kishimoto T. Cell reproduction: induction of M-phase events by cyclin–dependent cdc2 kinase. Int J Dev Biol 1994; 38:185–191.
16 Kranenburg O, van der Eb AJ, Zantema A. Cyclin–dependent kinases and pRb: regulators of the proliferation–differentiation switch. FEBS Lett 1995; 367:103–106.
17 Lees, E. Cyclin dependent kinase regulation. Curr Opin Cell Biol 1995; 7:773–780.
18 Lehner CF, Ried G, Stern B et al. Cyclins and cdc2 kinases in Drosophila: genetic analyses in a higher eukaryote. Ciba Found Symp 1992; 170:97–109; discussion 110–114.
19 Lehner CF. Pulling the string: cell cycle regulation during Drosophila development. Semin Cell Biol 1991; 2:223–231.
20 Maller JL. Mitotic control. Curr Opin Cell Biol 1991; 3:269–275.
21 Marcote MJ, Pagano M, Draetta G. cdc2 protein kinase: structure–function relationships. Ciba Found Symp 1992; 170:30–41; discussion 41–49.
22 Massague J, Polyak K. Mammalian antiproliferative signals and their targets. Curr Opin Genet Dev 1995; 5:91–96.
23 Masui Y. Towards understanding the control of the division cycle in animal cells. Biochem Cell Biol 1992; 70:920–945.
24 Minshull J. Cyclin synthesis: who needs it? Bioessays 1993; 15:149–155.
25 Morgan DO. Principles of CDK regulation. Nature 1995; 374:131–134.
26 Murray A. Cell cycle checkpoints. Curr Opin Cell Biol 1994; 6:872–876.
27 Murray A. Cyclin ubiquitination: the destructive end of mitosis. Cell 1995; 81:149–152.
28 Murray AW, Hunt T. The Cell Cycle: An Introduction. New York:WH Freeman and Co., 1993.
29 Murray AW. Creative blocks: cell cycle checkpoints and feedback controls. Nature 1992; 359:599–604.
30 Murray AW. The genetics of cell cycle checkpoints. Curr Opin Genet Dev 1995; 5:5–11.
31 Musgrove EA, Sutherland RL. Cell cycle control by steroid hormones. Semin Cancer Biol 1994; 5:381–389.
32 Nigg EA, Gallant P, Krek W. Regulation of $p34^{cdc2}$ protein kinase activity by phosphorylation and cyclin binding. Ciba Found Symp 1992; 170:72–84; discussion 84–96.
33 Nigg EA. Cyclin–dependent protein kinases: key regulators of the eukaryotic cell cycle. Bioessays 1995; 17:471–480.
34 Nigg EA. Targets of cyclin–dependent protein kinases. Curr Opin Cell Biol 1993; 5:187–193.
35 Nishimoto T, Uzawa S, Schlegel R. Mitotic checkpoints. Curr Opin Cell Biol 1992; 4:174–179.
36 Norbury C, Nurse P. Animal cell cycles and their control. Annu Rev Biochem 1992; 61:441–470.
37 Nurse P. Ordering S phase and M phase in the cell cycle. Cell 1994; 79:547–550.
38 O'Connell MJ, Nurse P. How cells know they are in G1 or G2. Curr Opin Cell Biol 1994; 6:867–871.
39 Peeper DS, van der Eb AJ, Zantema A. The G1/S cell–cycle checkpoint in eukaryotic cells. Biochim Biophys Acta 1994; 1198:215–230.

40 Peter M, Herskowitz I. Joining the complex: cyclin–dependent kinase inhibitory proteins and the cell cycle. Cell 1994; 79:181–184.
41 Pines J, Hunter T. Cyclins A and B1 in the human cell cycle. Ciba Found Symp 1992; 170:187-196; discussion 196–204.
42 Pines J. Cyclins and cyclin–dependent kinases: themes and variations. Adv Cancer Res 1995; 66:181–212.
43 Pines J. Cyclins and their associated cyclin–dependent kinases in the human cell cycle. Biochem Soc Trans 1993; 21:921–925.
44 Pines J. Protein kinases and cell cycle control. Semin Cell Biol 1994; 5:399–408.
45 Pines J. The cell cycle kinases. Semin Cancer Biol 1994; 5:305–313.
46 Poon RY, Hunter T. Cell regulation. Innocent bystanders or chosen collaborators? Curr Biol 1995; 5:1243–1247.
47 Reed SI, Bailly E, Dulic V et al. G1 control in mammalian cells. J Cell Sci Suppl 1994; 18:69–73.
48 Reed SI, Dulic V, Lew DJ et al. G1 control in yeast and animal cells. Ciba Found Symp 1992; 170:7–15; discussion 15–19.
49 Roberts JM, Koff A, Polyak K et al. Cyclins, Cdks, and cyclin kinase inhibitors. Cold Spring Harb Symp Quant 1994; 59:31–38.
50 Roberts JM. Turning DNA replication on and off. Curr Opin Cell Biol 1993; 5:201–206.
51 Satterwhite DJ, Moses HL. Mechanisms of transforming growth factor–beta–1–induced cell cycle arrest. Invasion Metastasis 1994–1995; 14:309–318.
52 Sherr CJ, Matsushime H, Roussel MF. Regulation of CYL/cyclin D genes by colony–stimulating factor 1. Ciba Found Symp 1992; 170:209–219; discussion 219–226.
53 Sherr CJ, Roberts JM. Inhibitors of mammalian G1 cyclin–dependent kinases. Genes Dev 1995; 9:1149–1163.
54 Sherr CJ. D-type cyclins. Trends Biochem Sci 1995; 20:187–190.
55 Sherr CJ. G1 phase progression: cycling on cue. Cell 1994; 79:551–555.
56 Sherr CJ. Growth factor–regulated G1 cyclins. Stem Cells (Dayt) 1994; 12:47-55; discussion 55–57.
57 Solomon MJ. Activation of the various cyclin/cdc2 protein kinases. Curr Opin Cell Biol 1993; 5:180–186.
58 Stein GH, Dulic V. Origins of G1 arrest in senescent human fibroblasts. Bioessays 1995; 17:537–543.
59 Vousden KH. Regulation of the cell cycle by viral oncoproteins. Semin Cancer Biol 1995; 6:109–116.
60 Yanagida M, Kinoshita N, Stone EM et al. Protein phosphatases and cell division cycle control. Ciba Found Symp 1992; 170:130–140; discussion 140–146.
61 Zetterberg A, Larsson O, Wiman KG. What is the restriction point? Curr Opin Cell Biol 1995; 7:835–842.

Yeast Cell Cycle Genetics

1 Cross FR. Starting the cell cycle: what's the point? Curr Opin Cell Biol 1995; 7:790–797.
2 D'Urso G, Nurse P. Checkpoints in the cell cycle of fission yeast. Curr Opin Genet Dev 1995; 5:12–16.
3 Fisher D, Nurse P. Cyclins of the fission yeast Schizosaccharomyces pombe. Semin Cell Biol 1995; 6:73–78.
4 Fleig UN, Gould KL. Regulation of cdc2 activity in Schizosaccharomyces pombe: the role of phosphorylation. Semin Cell Biol 1991; 2:195–204.
5 Forsburg SL, Nurse P. Cell cycle regulation in the yeasts Saccharomyces cerevisiae and Schizosaccharomyces pombe. Annu Rev Cell Biol 1991; 7:227–256.
6 Futcher AB. Saccharomyces cerevisiae cell cycle: cdc28 and the G1 cyclins. Semin Cell Biol 1991; 2:205–212.

7 Goebl MG, Winey M. The yeast cell cycle. Curr Opin Cell Biol 1991; 3:242–246.
8 Millar JB, Lenaers G, McGowan C et al. Activation of MPF in fission yeast. Ciba Found Symp 1992; 170:50–58; discussion 58–71.
9 Nasmyth K. Control of the yeast cell cycle by the Cdc28 protein kinase. Curr Opin Cell Biol 1993; 5:166–179.
10 Sherlock G, Rosamond J. Starting to cycle: G1 controls regulating cell division in budding yeast. J Gen Microbiol 1993; 139:2531–2541.

Tumor Suppressor Genes

1 Cobrinik D, Dowdy SF, Hinds PW et al. The retinoblastoma protein and the regulation of cell cycling. Trends Biochem Sci 1992; 17:312–315.
2 Ewen ME, Miller SJ. p53 and translational control. Biochim Biophys Acta, Gene Struct Expr 1996; 1242:181–184.
3 Ewen ME. The cell cycle and the retinoblastoma protein family. Cancer Metastasis Rev 1994; 13:45–66.
4 Haffner R, Oren M. Biochemical properties and biological effects of p53. Curr Opin Genet Dev 1995; 5:84–90.
5 Hainaut P. The tumor suppressor protein p53: a receptor to genotoxic stress that controls cell growth and survival. Curr Opin Oncol 1995; 7:76–82.
6 Hinds PW, Weinberg RA. Tumor suppressor genes. Curr Opin Genet Dev 1994; 4:135–141.
7 Hinds PW. The retinoblastoma tumor suppressor gene. Curr Opin Genet Dev 1995; 5:79–83.
8 Hollingsworth RE Jr, Hensey CE, Lee WH. Retinoblastoma protein and the cell cycle. Curr Opin Genet Dev 1993; 3:55–62.
9 Weinberg RA. The retinoblastoma protein and cell cycle control. Cell 1995; 81:323–330.
10 Whyte P. The retinoblastoma protein and its relatives. Semin Cancer Biol 1995; 6:83–90.
11 Zacksenhaus E, Bremner R, Jiang Z et al. Unraveling the function of the retinoblastoma gene. Adv Cancer Res 1993; 61:115–141.
12 Zhu L, Enders GH, Wu CL et al. Growth suppression by members of the retinoblastoma protein family. Cold Spring Harb Symp Quant Biol 1994; 59:75–84.

Transcriptional Control

1 Adams PD, Kaelin WG. Transcriptional control by E2F. Semin Cancer Biol 1995; 6:99–108.
2 Cross F. Transcriptional regulation by a cyclin cdk. Trends Genet 1995; 11:209–211.
3 Kouzarides T. Transcriptional control by the retinoblastoma protein. Semin Cancer Biol 1995; 6:91–98.
4 La Thangue NB. DP and E2F proteins: components of a heterodimeric transcription factor implicated in cell cycle control. Curr Opin Cell Biol 1994; 6:443–450.
5 La Thangue NB. DRTF1/E2F: an expanding family of heterodimeric transcription factors implicated in cell–cycle control. Trends Biochem Sci 1994; 19:108–114.
6 Lam EW, La Thangue NB. DP and E2F proteins: coordinating transcription with cell cycle progression. Curr Opin Cell Biol 1994; 6:859–866.
7 Muller R. Transcriptional regulation during the mammalian cell cycle. Trends Genet 1995; 11:173–178.
8 Nevins JR, Chellappan SP, Mudryj M et al. E2F transcription factor is a target for the RB protein and the cyclin A protein. Cold Spring Harb Symp Quant Biol 1991; 56:157–162.
9 Seroz T, Hwang JR, Moncollin V et al. TFIIH: a link between transcription, DNA repair and cell cycle regulation. Curr Opin Genet Dev 1995; 5:217–221.
10 Weinberg RA. E2F and cell proliferation: a world turned upside down. Cell 1996; 85:457–459.

Chapter 2

Discovery of Cell Cycle Regulators

The cell theory, proposed by Schleiden and Schwann in 1838, is the foundation on which modern biology rests. Simply stated, the theory says that all forms of earthly life are ultimately composed of cells, and all cells arise from the division of preexisting cells. It is the division of cells that concerns us in this book. Controlled division of cells is crucial to every biological process from development to reproduction. The uncontrolled division of cells, of course, can lead to cancer.

Cytological studies in the early 1900s detailed the processes of mitosis and cell division, although half a century would pass before we conclusively identified and characterized the genetic material of deoxyribonucleic acid. Shortly thereafter we could explain in molecular terms the complex set of events underlying cell division, from DNA replication to chromosome condensation to the formation of the spindle apparatus to cytokinesis. The mechanics of cell division were readily accessible to us; nevertheless, the regulatory scheme that ordered these events remained an impenetrable mystery. Only in the past couple of decades have we developed sufficiently sophisticated genetic and biochemical techniques to probe the innermost workings of the cell. The information that has come to light has rapidly advanced our understanding of cell division or, more broadly speaking, the cell cycle.

The generally accepted formulation of the eukaryotic cell cycle recognizes the two most significant sets of events in a cell's lifespan, the replication of the nuclear DNA and the dual proliferative processes of mitosis (or meiosis) and cytokinesis. The former takes place during the S (synthesis) phase, and the latter is restricted to the M (mitosis) phase. Juxtaposed between these two phases are the gap phases: the G1 phase follows M phase and precedes S phase; the G2

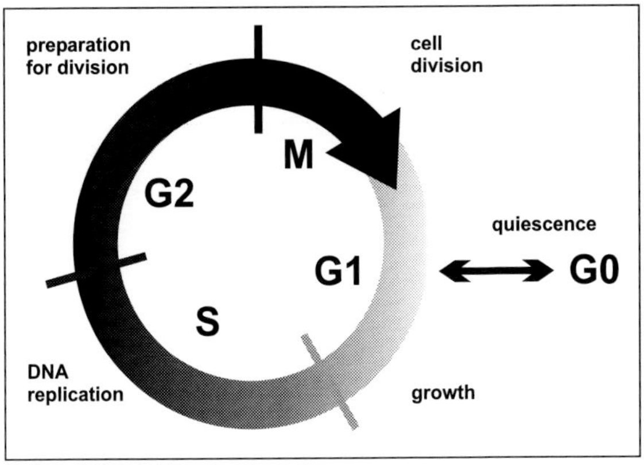

Fig. 2.1. Diagram depicting the cell (division) cycle of a typical rapidly–dividing mammalian cell. The two most significant sets of events in a cell's lifespan are DNA replication and cell division, which occur in S phase and M phase respectively. Juxtaposed between the S and M phases are the two gap phases, G1 and G2. G1 is a period distinguished by cell growth, and G2 has been described as an interval in which the cell prepares for division. G0 phase, in which the cell abides in a quiescent, non–growing state, can be regarded as "outside" the cell cycle; entry into and exit from G0 occurs at a point during G1. This diagram roughly illustrates the size of the cell (darker signifies larger) as it traverses the cell cycle, as well as the relative lengths of the phases.

phase follows S phase and precedes M phase. In typical rapidly–dividing mammalian cells, which are the subject of primary interest in this book, approximately 5% of the cycle is spent in M phase, whereas the rest is spent in the other three phases, collectively called interphase. Cells can also abide for an indefinite period of time in a quiescent state designated G0 phase, which can be thought of as a suspension of the cell cycle (Fig. 2.1).[1]

The arbitrary division of the cell cycle into these phases (some types of cells do not adhere to this schematic, the notable exceptions being meiotic and embryonic cells) begs the question of how a cell is informed of what phase it currently resides in, and what determines whether it should make the transition into the next chronological phase. These are issues currently under investigation, by no means close to resolution. The emerging evidence, however, suggests that proteins known as cyclins and cyclin–dependent kinases (CDKs) may be the crucial indicators and determinants of the cell's position in the cycle. It is worth reviewing how these molecules were discovered before we examine the various cyclins and CDKs in some depth.

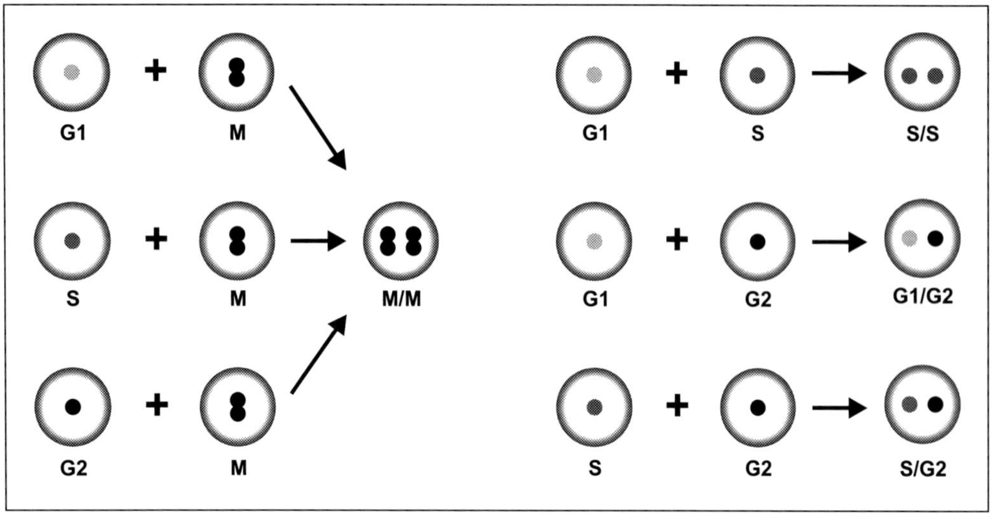

Fig. 2.2. Diagram depicting the results of a series of mammalian cell fusion experiments published in 1970.[2,3] Fusing any interphase cell with an M phase cell yields a hybrid with two M phase nuclei; fusing a G1 cell with an S cell yields a hybrid with two S phase nuclei; fusing a G1 and G2 cell, or an S cell and G2 cell, leaves the hybrid nuclei unchanged from their original states. These data indicate the existence of a "mitosis–promoting factor" during M phase and a "synthesis–promoting factor" during S phase, as well as a "synthesis–inhibiting factor" during G2 phase.

The Structure of the Cell Cycle

A series of mammalian cell fusion experiments published in 1970 revealed that there is an underlying structure to the cell cycle.[2,3] When a cell in any part of interphase was fused with a cell in M phase, the interphase nucleus displayed chromosomal condensation, an event usually associated with a mitotic nucleus. When a cell in G1 phase was fused with a cell in S phase, the G1 nucleus initiated replication of its DNA as if it were in S phase. Moreover, mitosis did not commence until after both nuclei had completed their DNA replication and "progressed" past S phase. When an S phase cell was fused with a G2 phase cell, neither nucleus was affected, but the onset of mitosis was delayed until the S nucleus had completed its replication. Similarly, fusing a G1 phase cell to a G2 phase cell did not affect the G1 nucleus' progression through the cell cycle, although the G2 nucleus was prevented from entering mitosis until the other nucleus had progressed into G2 (Fig. 2.2).

The experiments suggest the following conclusions. M phase cells contain an activator protein or proteins that interphase cells lack; when interphase nuclei are exposed to this "mitosis–promoting factor," they become mitotic nuclei. S phase cells similarly uniquely contain a "synthesis–promoting factor" that converts G1 nuclei into S nuclei. G2 nuclei, however, are immune to this synthesis–promoting factor, for they remain unchanged instead of undergoing another round of DNA replication as might be expected. The G2 nucleus evidently has some mechanism that protects its DNA from further replication until after mitosis has occurred. Finally, mitosis does not commence until all the nuclear DNA in the cell has undergone complete replication.

The lesson of these experiments is that there is complex machinery at work in the cell cycle. There are conditions to be satisfied, factors to be produced, and inhibitory mechanisms to be disengaged before the cell can progress from phase to phase. It is from these experiments and subsequent work that the concept of "checkpoints" has evolved.[4,5] Cells pause at various checkpoints before crucial phase transitions to interpret internal and environmental conditions via feedback controls; if upon synthesizing the available information the cell determines conditions are favorable, it will proceed to the next part of the cycle. Unfavorable circumstances cause the cell to be arrested at the checkpoint. The advantages of checkpoint control are obvious. It allows an organism to tightly regulate the proliferative level of its cells via extracellular signals. Individual cells will normally not reproduce unless all the proper signals have been received. Checkpoints being improperly overridden is a prerequisite for oncogenesis; we must therefore elucidate the mechanisms that regulate the checkpoints if we are to understand cancer.

For cells to be self–monitoring requires a high level of sophistication, presumably involving a large number of molecules interacting in a variety of interconnected pathways. The first of these regulatory molecules to be characterized were the cyclins.

The First Cyclins

The existence of a "mitosis–promoting factor" unique to M phase and a "synthesis–promoting factor" unique to S phase aroused suspicions that phase–specific proteins were periodically synthesized from cycle to cycle and that these proteins drove the cell through the phase transitions. However, the search for such "periodic" proteins turned up no candidates. A study published in 1983 unexpectedly found proteins with unusual cyclic behavior in the eggs of two species of sea urchin, *Arbacia punctulata* and *Lytechinus pictus*, and a species of surf clam, *Spisula solidissima*.[6] When eggs of *A. punctulata* were fertilized or chemically activated, the synthesis of three new proteins was observed. Two of these proteins increased steadily in quantity during the first few cell cycles, reflecting

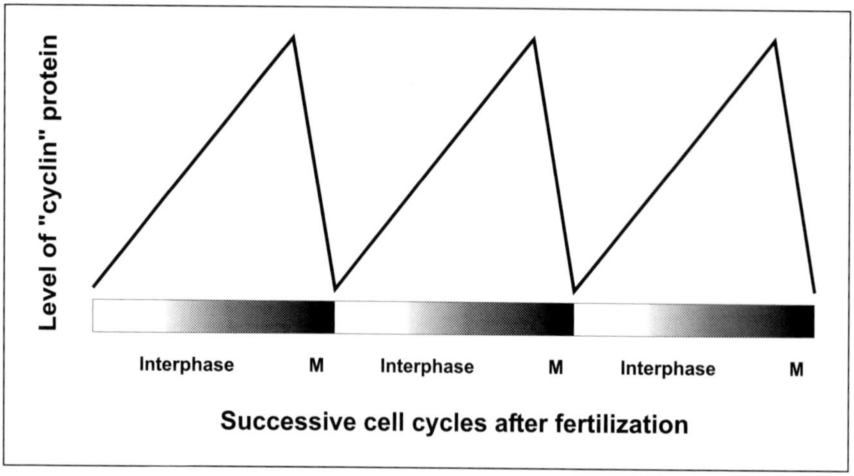

Fig. 2.3. Diagram depicting the behavior of the "cyclin" protein in successive cell cycles following the fertilization of eggs of the sea urchin species *Arbacia punctulata*.[6] The protein undergoes gradual synthesis during interphase followed by rapid destruction during M phase.

their continuous synthesis. In contrast, the third protein increased steadily during each interphase but suddenly declined at each cleavage division. This protein, dubbed "cyclin," was shown to undergo continuous synthesis punctuated by periodic depletion (Fig. 2.3).

Analogous experiments performed with eggs of *L. pictus* and *S. solidissima* found two proteins with cyclin–like behavior.[6] Significantly, the proteins did not disappear and reappear in synchrony but oscillated out of phase, peaking at slightly different times. As a result of their distinct behaviors, and because the proteins had previously been reported (although not in the context of cell cycle–dependent behavior) in *S. Solidissima* and designated proteins A and B,[7] the two proteins came to be known as cyclin A and cyclin B (Fig. 2.4). The peak levels of the cyclins coincided with the start of M phase, suggesting that at least one played a role in driving cells into mitosis. The demonstration that the injection of clam cyclin A mRNA into oocytes of the frog species *Xenopus laevis* caused the onset of meiosis led to speculation that cyclin A was a primary mitotic inducer,[8] although this turned out to be close to but not quite the truth (see next section).

Investigators were confronted with "periodic proteins" that were periodic not in their synthesis, but rather in their destruction. Assuming the cyclins were involved in the onset of mitosis, one could sketch out the following regulatory scheme. Cells would accumulate cyclin throughout the cell cycle. When the quan-

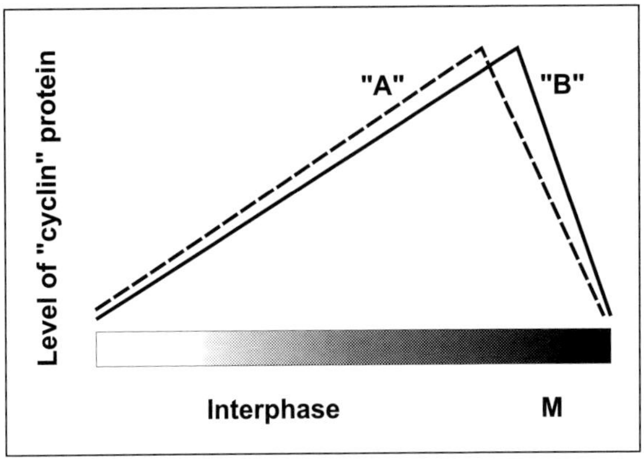

Fig. 2.4. Diagram depicting the behavior of two proteins during the cell cycle of the eggs of the sea urchin species *Lytechinus pictus* and the surf clam species *Spisula solidissima*.[6] Both display the periodic pattern of continuous synthesis punctuated by periodic depletion, but they peak at slightly different times near the start of M phase. These proteins are now known as cyclin A and cyclin B.

tity of cyclin surpassed a threshold level, mitosis would commence. During mitosis, the cyclin would be destroyed. Only by eliminating the store of cyclin could the cell assure that the next mitosis would occur after a proper interval. Numerous studies with a variety of eukaryotic cells had established that some protein or proteins must be synthesized to permit entry into M phase by demonstrating that a chemical–induced block of protein synthesis inhibits mitotic events.[9-18] The requirement for protein synthesis is conveniently explained if cyclin must be resynthesized every cycle in sufficient quantity to surpass the threshold and effect mitosis.

The simple regulatory model we have outlined is notable for its vulnerability: if through mutation the expression of cyclin were amplified, or if the mechanism responsible for destroying the cyclin were faulty, the resulting excess of cyclin would promote premature mitoses, subjecting cells to accelerated proliferation. In such cells would reside the potential for oncogenesis. Fortunately (for our well–being) the regulation of the cell cycle is far more complicated than we have just supposed. More recent work has demonstrated that the role of the cyclins is only one of a number of controls on the onset of mitosis, which combine to preclude uncontrolled cell division.

Mitosis–Promoting Factor

At roughly the same time the cell fusion experiments discussed earlier in the chapter postulated the existence of "mitotis–promoting factor," studies with eggs of the frog species *Rana pipiens* discovered that this factor was indeed a

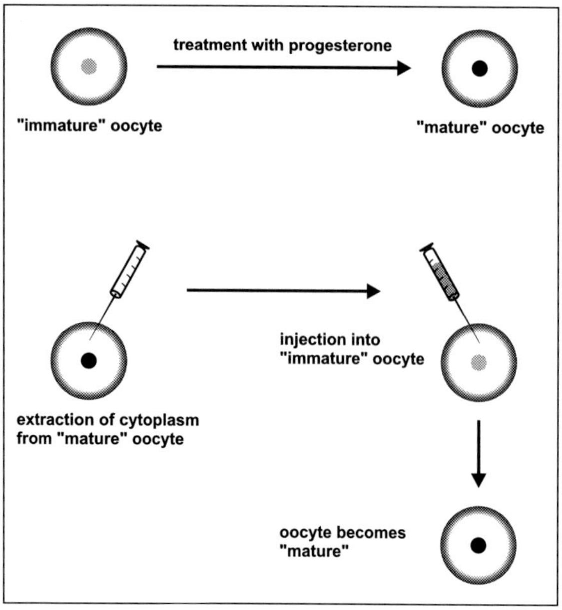

Fig. 2.5. Diagram depicting the results of experiments with the eggs of the frog species *Rana pipiens.*[19-21] Treating "immature" oocytes—i.e., unfertilized ooctyes arrested in first meiotic prophase—with progesterone causes them to become "mature," advancing to second meiotic metaphase. Extracting a cytoplasmic sample from a mature oocyte and injecting it into an immature cell causes the latter to become mature. This result suggests the existence of a "maturation–promoting factor" present only in mature cells.

discrete entity.[19-21] Fully grown, unfertilized *R. pipiens* oocytes are arrested in first meiotic prophase; treatment with hormones such as progesterone cause the oocytes to "mature" to second meiotic metaphase. In the experiments, cytoplasm withdrawn from progesterone–treated oocytes and injected into non–treated oocytes caused the latter to mature (Fig. 2.5), whereas cytoplasm withdrawn from non–treated cells could not duplicate the effect (Fig. 2.6). Nor could progesterone injected directly into oocytes induce maturation. Clearly progesterone caused the activation of a cytoplasmic factor that could drive cells through meiosis. Moreover, the factor was produced in oocytes whose nuclei were removed prior to progesterone treatment. This indicated that the factor (or components of the factor) were present in the cytoplasm in inactive form before the induction of maturation and that progesterone treatment activated the factor. In retrospect, this was a significant clue as to the molecular identity of the factor.

Designated maturation–promoting factor (and later, mitosis–promoting factor) or MPF, the cytoplasmic factor was subsequently found in cells from a host

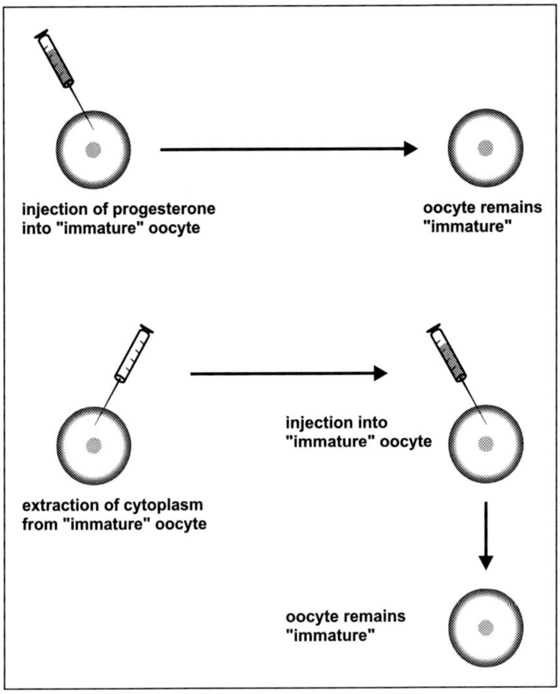

Fig. 2.6. Diagram depicting the control experiments for the studies described in Figure 2.5. Directly injecting progesterone into an immature oocyte does *not* cause the cell to become mature, in contrast with the external treatment of the oocyte with the hormone. This indicates that progesterone is not the "maturation–promoting factor" itself but rather induces the production of the factor when exogenously presented to immature cells. Moreover, a cytoplasmic extract from an immature oocyte is incapable of inducing maturation when injected into another immature cell, establishing that the maturation process is not triggered by the mechanical process of extraction and injection but by a genuine cytoplasmic factor present only in mature oocytes.

of other eukaryotes[22-31] by means of an assay in which cytoplasmic extracts from these cells were tested for their ability to induce maturation of frog oocytes.[32,33] MPF was also shown to be able to induce in embryonic extracts a host of mitotic events, including chromosome condensation, disintegration of the nuclear membrane, and the formation of the mitotic spindle apparatus.[34,35] Further, MPF activity was seen to fluctuate during the cell cycle, reaching a peak during M phase.[23,24,36] Concurrent work had demonstrated the existence of a protein kinase, detected by its ability to phosphorylate histone proteins, whose activity oscillated during the cell cycle and was maximal in M phase.[37-43] This behavior earned the entity the name of M phase histone H1 kinase. Not surprisingly, there was much speculation that the kinase activity and MPF were functionally equivalent; the logical extension was that MPF in part comprised the protein kinase.

The identification of the components of MPF eluded investigators until it was purified from *X. laevis* and starfish eggs.[44-46] Analysis of the factor revealed it to contain two protein species, one of 45-47 kDa and one of 32-34 kDa in apparent molecular weight (when separated on a polyacrylamide gel). The larger protein was quickly identified as a B-type cyclin, confirming the long–suspected connection between cyclins and mitosis–promoting factor.[47,48] Notably, cyclin B has no kinase domain and no known enzymatic function. It was therefore concluded that cyclin B was not the origin of the observed kinase activity, but rather played some sort of regulatory role with respect to the kinase.

It was immediately evident that the smaller protein species in MPF was the suspected protein kinase. As it turned out, a separate line of genetic experimentation years in the making had already implicated a previously identified kinase as being associated with MPF. The history of this kinase deserves examination, because it was only when it was definitively shown that the kinase was the catalytic subunit of MPF did investigators realize that they had in hand the first of the cyclin–dependent kinases.

Cell Division Cycle Genes

The cell division cycle (CDC) genes were first identified in yeast cells of the budding type *Saccharomyces cerevisiae* and the fission type *Schizosaccharomyces pombe* (Fig. 2.7).[49-54] Out of necessity they were isolated as conditional mutants that in nonpermissive conditions caused the cells to arrest in various phases of the cell cycle yet still grow in size (nonconditional mutants would have been undetectable since the cells would not have proliferated). Thus the genes' products were shown to be essential specifically for the progression of the phases of the cell cycle rather than, for instance, the biosynthetic pathway of some important molecule, which if blocked would also cause cessation of the cell cycle.[55] CDC genes have also been isolated by their ability, when overexpressed or mutated, to suppress or complement the effects of another CDC gene. To date more than 50 *S. cerevisiae* and 35 *S. pombe* CDC genes have been identified.[1] (Fig. 2.7 shows a selection of CDC genes with their suspected points of action during the cell cycle of the appropriate organism.) From this large pool of genes investigators have found several cell cycle regulators that are conserved across the species. The two most noteworthy are *cdc2⁺/CDC28* and *cdc13⁺* (the names in lower case letters designate *S. pombe* genes, the names in upper case letters, *S. cerevisiae* genes).

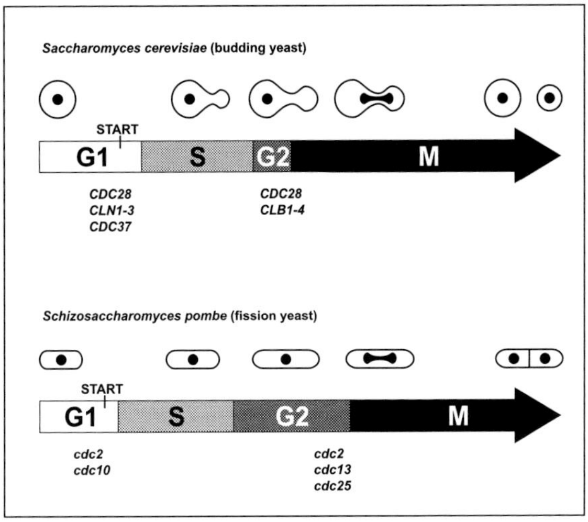

Fig. 2.7. Diagrams depicting the cell cycles of the budding yeast *Saccharomyces cerevisiae* and the fission yeast *Schizosaccharomyces pombe*. *S. cerevisiae* cells proliferate by forming a bud, replicating the genome, moving one set of genes into the bud, and separating the "daughter" bud from the "mother" cell in an unequal division. *S. cerevisiae* is notable for its very brief or non–existent G2 phase and lengthy M phase. In contrast, *S. pombe* cells reproduce by growing larger (longer), replicating the genome, dividing the two sets of genes and dividing into two equal–sized progeny. Shown on the diagrams are the START checkpoints, the decision points which once passed commit the cells to completing the remainder of their cell cycles. Also shown are two select groups of CDC and related genes which, when mutated, cause the cells to arrest either before the G1/S phase transition or the G2/M transition. It must be noted that the cell cycles presented here are not the only possible fates for the yeasts, but only one of several alternatives including mating with other cells and sporulation.

cdc2⁺/CDC28 was characterized in connection to START, a cell cycle checkpoint prior to the G1/S phase transition (Fig. 2.7).[50,56-60] In *S. cerevisiae* and *S. pombe*, the regulation of START is the single most important determinant of whether a cell will commit to another cycle; while in G1 phase, the yeast cell can choose among the diverging fates of mitosis, mating with another cell, or sporulation; once S phase is entered, the cell is required to execute DNA synthesis and mitosis. *CDC28* was originally identified as a START mutant; budding yeast defective in this gene arrested late in G1 phase.[56,61] *CDC28* was also implicated in causing a block in G2 phase.[62,63] Similarly, deletion of the *cdc2*$^+$ gene in fission yeast caused cells to be halted at two points, before either the G1/S or G2/M phase transition.[58,64] Also, mutational analysis of *cdc2*$^+$ revealed that its gene product was a primary determinant of the length of G2 phase.[64,65] When it was discovered that *CDC28* could rescue *cdc2*–deficient fission yeast from the arrested phenotype when introduced into the cells, and *cdc2*$^+$ could do the same for *CDC28*–deficient budding yeast, and that the two yeast genes were homologues with respect to their nucleotide sequences, it was clear that investigators had found two versions of a single protein essential to cell cycle control.[66,67] The universality of this protein was confirmed within a few years by the identification of homologues of *cdc2*$^+$/*CDC28* in a wide variety of eukaryotes, including fruit flies, plants, mice, chickens, and humans.[68-79]

cdc2$^+$/*CDC28* was found to encode a 34 kDa serine/threonine protein kinase, designated $p34^{cdc2}$.[80-83] As a kinase, $p34^{cdc2}$ held the potential to phosphorylate and thus activate or inactivate multiple downstream targets, certainly consistent with its putative role as a key regulator of two cell cycle checkpoints. To ensure the kinase would not act at inappropriate times, however, there must have existed regulatory mechanisms embodied in proteins that interacted with $p34^{cdc2}$.

One such protein was identified by its ability to suppress the deleterious effect of $p34^{cdc2}$ mutants. The *cdc13*$^+$ gene had previously been found to cause cell cycle arrest at the G2/M phase transition when deleted in *S. pombe*, implying a role in the induction of mitosis. Certain mutants of *cdc13* were shown to counteract the effects of one partially defective allele of *cdc2*, and overexpressed *cdc13*$^+$ was able to rescue other conditional *cdc2* mutants.[54,84,85] This observation initially led to speculation that the product of *cdc13*$^+$ was a substrate of $p34^{cdc2}$.[85] Upon cloning and sequencing *cdc13*$^+$, however, investigators discovered that this cell division cycle gene was homologous to B–type cyclins.[86-88] Later work showed that B–type cyclins associated with $p34^{cdc2}$ function existed in *S. cerevisiae* as well.[89,90]

In a short span of time the various biochemical and genetic lines of evidence we have documented here converged to a single, elegant description of cell cycle regulation.[91] Levels of cdc13 protein were seen to decline abruptly during mitosis in yeast cells, exactly as one would expect of a B–type cyclin.[92,93] It was observed that $p34^{cdc2}$ and cdc13 protein could physically associate in a stable complex, and that in yeast cells both proteins were localized in the nucleus during mitosis; moreover, $p34^{cdc2}$–associated kinase activity dropped as soon as cdc13 was depleted, and $p34^{cdc2}$ disappeared from the nucleus.[93] The clincher was that $p34^{cdc2}$ was the same size as the smaller protein species isolated from purified MPF, which was 32-34 kDa. Indeed, the protein kinase activities imputed to MPF and the M phase histone H1 kinase were both shown to be none other than $p34^{cdc2}$.[43,47,94,95] Shortly thereafter came confirmation that $p34^{cdc2}$ was an essential mitotic inducer in eukaryotes other than yeast, including mice and humans.[96,97] Along with the observation that cyclin B was a component of MPF,[47,48] the evidence pointed to an inescapable conclusion: MPF is a cyclin B–$p34^{cdc2}$ complex conserved among eukaryotic species.

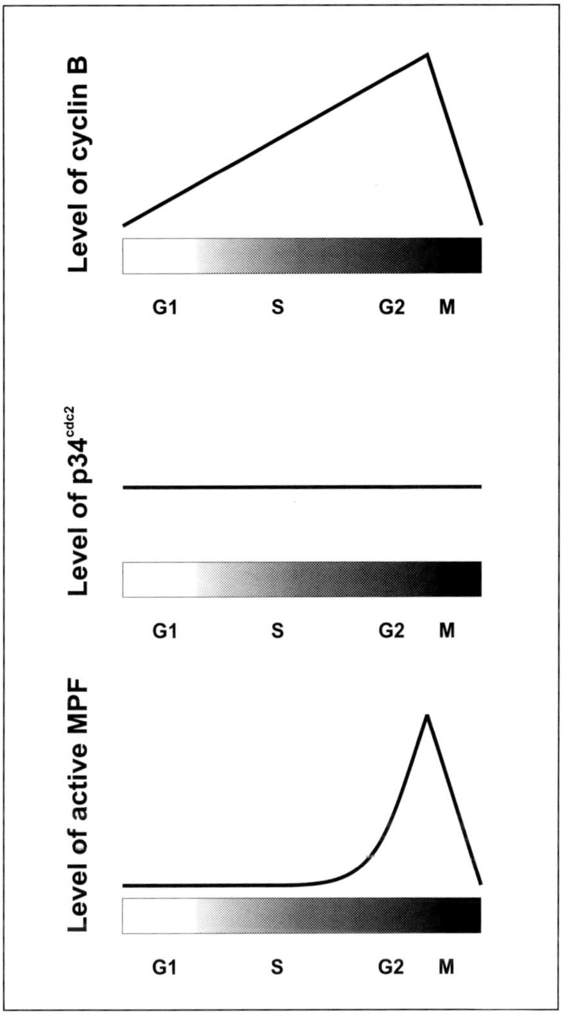

Fig. 2.8. Diagrams depicting the levels of cyclin B, p34cdc2, and mitosis–promoting factor (MPF) during the cell cycle. Whereas cyclin B slowly accumulates during the G1, S and G2 phases and peaks at the G2/M phase transition, subsequently undergoing rapid degradation in M phase, p34cdc2 remains at a roughly constant level throughout all the phases. Active MPF is detected during a limited interval of the cell cycle, rising sharply in late G2 and precipitously dropping in M. These observations are consistent with a model in which cyclin B and p34cdc2 are components of MPF and are both required for the factor to assume kinase activity. It is only in G2 that the cyclin B level reaches the stoichiometric threshold past which the cyclin can bind to p34cdc2 to form active MPF complexes. In M phase, the destruction of cyclin B breaks up the MPF complexes, corresponding to a sharp drop in activity.

MPF is a Cyclin–CDK Complex

The identification of cyclin B and p34cdc2 was the culmination of years of work that enabled investigators to sketch out a new model of the regulation of the mitotic checkpoint (Figs. 2.8, 2.9). In contrast to cyclin B, with its periodic accumulation and destruction, p34cdc2 was maintained in constant quantity throughout the cell cycle. By itself the kinase was inactive; only when the cyclin level was high enough would cyclin B and p34cdc2 assemble into an MPF complex. As part of the complex, p34cdc2 was an active catalytic subunit that would

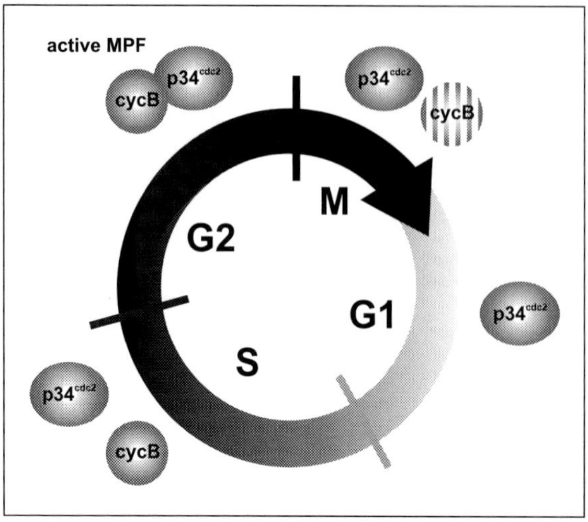

Fig. 2.9. Schematic illustration of the model of formation and dissolution of MPF complexes during the cell cycle. In G1 phase, only $p34^{cdc2}$ is present in appreciable quantities. By S phase, a sizable amount of cyclin B has been synthesized, but it is only in G2 phase that there is sufficient cyclin to bind $p34^{cdc2}$ into active MPF complexes. MPF catalyzes the cell's entry into mitosis, at which time cyclin B is degraded, eliminating the MPF complexes. The cell cycle begins anew with the presence of $p34^{cdc2}$ and absence of cyclin B.

phosphorylate multiple downstream targets and thus initiate mitotic events. Since cyclin B peaked at the end of G2 phase, it was only during this part of the cell cycle that $p34^{cdc2}$ could be a functional kinase. The degradation of the cyclin B subunit during mitosis unraveled the MPF complex, eliminating the kinase activity associated with $p34^{cdc2}$. This prevented the kinase from inducing another set of mitotic events immediately after the completion of the first mitosis.

Cyclin B, then, was the regulatory subunit of MPF. Whether it operated by binding $p34^{cdc2}$ and ferrying it to the nucleus where the key substrates (with respect to mitotic induction) of the kinase were located, or by modifying the conformation of the kinase to activate it, or both, was unclear. The first alternative was supported by the finding that $p34^{cdc2}$ was located in the nucleus during only mitosis in yeast cells;[93] but the issue was clouded by the early observation that the kinase remained in the nucleus throughout the cell cycle in mammalian cells,[96] although recent studies have demonstrated that human cyclin B indeed localizes to cytoplasmic structures during interphase.[98,99] The second alternative was reinforced with evidence of cyclin B altering the catalytic properties of $p34^{cdc2}$.[93,100,101]

Regardless of how cyclin B exerted its influence on $p34^{cdc2}$, it was clear that the latter's activity was dependent on a physical interaction with the former.

Because of this novel regulatory scheme, p34^{cdc2} came to be known as a cyclin–dependent kinase (CDK). Within a short time after the characterization of MPF a number of other CDKs were discovered, and p34^{cdc2} is now identified as CDK1, a name by which we will refer to this protein kinase for the remainder of this text to avoid confusion.

Another Cyclin, Another CDK

Concurrent with the discovery that MPF was a complex of cyclin B and CDK1 was the finding that cyclin A could also bind to CDK1.[100-102] However, it became apparent that the complexes of cyclin A-CDK1 and cyclin B-CDK1 had different substrate specificities.[103] Further, while a number of cyclin B homologues were identified in the *S. cerevisiae* and *S. pombe* systems, no distinct cyclin A homologue was found in the yeasts, suggesting that cyclin A played a specialized role distinct from that of cyclin B in the regulatory schemes of eukaryotic cells.

Most interestingly, cyclin A was found to associate with a protein closely related to CDK1. Discovered in *X. laevis*, *Drosophila* fruit flies and humans, the protein was found to be of similar size to CDK1 (33 kDa vs. 34 kDa) and to share significant homology in the amino acid sequence, for which reason it was designated p33^{cdk2} or CDK2.[76,94,104-110] Cyclin A but not cyclin B could form complexes with CDK2, confirming that A–type cyclins possessed regulatory functions distinct from those belonging to B–type cyclins.[111] CDK2 activity was observed to be high through much of interphase, in contrast to CDK1 activity, which was high only at the start of mitosis; moreover, both cyclin A and CDK2 activity were shown to be necessary for DNA replication in S phase.[112,113] The implication of these observations was that cyclin A functioned in two capacities during the cell cycle: when complexed with CDK1 it played a role in mitotic induction (albeit a different role from cyclin B–CDK1), and when complexed with CDK2 it participated in the regulation of DNA synthesis.

Cyclins Regulate the G1/S Phase Transition

In the yeast systems of *S. cerevisiae* and *S. pombe* two important checkpoints in the cell cycle had been identified: START in late G1 phase, and the G2/M phase transition. It had been convincingly shown that CDK1, as part of the MPF complex, was the regulator of the so–called mitotic checkpoint. But the *S. cerevisiae* version of the gene encoding CDK1, *CDC28*, had originally been identified because of its connection to START. Furthermore, the human homologue of CDK1 had been found to be able to substitute for *cdc2⁺* in *S. pombe* at either the G1/S or G2/M transition.[69] Not surprisingly, investigators wondered if a cyclin–CDK complex similar to MPF might be responsible for the regulation of START, i.e., act as an "SPF" (synthesis–promoting factor). Since cyclin B reached its maximal level at the beginning of M phase, and cyclin A peaked slightly prior to cyclin B, it would be a hitherto undiscovered cyclin that paired with CDK1 to form an active SPF complex and drive the yeast cell through START.

Two candidate proteins were identified by their ability to suppress conditional mutants of *CDC28* and failure to suppress deletion *CDC28* mutants, suggesting that they cooperated with and were dependent on $p34^{cdc2}$ to fulfill their roles. Designated *CLN1* and *CLN2*, the proteins had sequence homology to cyclins although they did not perfectly correspond to either A–type or B–type cyclins.[114] A third protein *CLN3*, alternatively known as *WHI1–1* or *DAF1–1*, had been isolated years before as a mutant that caused a shortening of G1 phase[115,116] and reidentified in a subsequent screen for G1 mutants.[117] *CLN3* was also found to be homologous to cyclins.[118]

Elegant experimentation revealed the importance of *CLN1*, *CLN2* and *CLN3* in the *S. cerevisiae* cell cycle. When any one or combination of two of the cyclin–like proteins were deleted, cells sustained a lengthened G1 phase and the consequent increase in size but otherwise grew normally; when all three proteins were deleted, cells suffered a G1 phase arrest, causing a lethal phenotype.[119,120] Thus the three proteins were redundant for some essential function in G1 phase. *CLN1*, *CLN2* and *CLN3* were categorized as "G1 cyclins," with the attendant speculation that they formed SPF complexes with CDK1, pushed the cell past the START checkpoint, and initiated DNA synthesis. This view received justification with the observation that *CLN2* associated with CDK1 to form an active protein kinase complex.[121]

The triple–*CLN* mutant emerged as an immediately useful system with which to study cyclin function in other species, for any protein that could rescue the lethal phenotype was almost certainly a cyclin, potentially a G1 cyclin. One research group used the mutant yeast to isolate human homologues of the *S. cerevisiae* G1 cyclins.[122] By a number of other means, other groups identified additional homologues in *S. pombe*, mice, and humans.[123-127] Their work led to the classification of three new classes of mammalian cyclins: C–type, D–type and E–type. An array of observations later demonstrated that cyclins of at least two of these classes associated with CDKs to form protein kinase complexes important in regulating the restriction (R) point, the late G1 phase checkpoint in animal cells analogous to START in yeast,[128] and the initiation of DNA replication. This is a subject that will be examined in greater detail in the next chapter.

A Paradigm of Cell Cycle Regulation

With the knowledge gleaned from 20 years of cell cycle research, investigators had arrived at a new paradigm of cell cycle regulation: checkpoint control by cyclin–CDK complexes. The induction of mitosis was clearly under the jurisdiction of MPF in the form of cyclin B–CDK1, and it seemed likely that DNA synthesis was contingent on the action of other cyclin–CDK factors. It was conceivable that every crucial decision for a cell to proceed to the next part of the cell cycle was executed by such complexes.

While the identification of cyclins and CDKs as the primary regulators of the cell cycle was nothing less than a scientific breakthrough, by no means did it give us the entire story. A host of details remained (and remain) to be elucidated. Even before cyclins were discovered investigators had been amassing evidence that CDK1 was subject to regulation by phosphorylation by other protein kinases. By the time the identity of MPF was unmasked, it was clear that these other kinases played an important role in activating the cyclin–CDK complex (a subject to be taken up in chapter 4). It was discovered that not only did the degradation of MPF take place in late mitosis, it was an event that was required to occur before mitosis could come to an end. More recently, a new group of cell cycle regulators have been identified and their inhibitory interaction with cyclins and CDKs clarified. And some of the signal transduction pathways that feed information to the cyclins–CDKs, as well as the targets of these complexes, have become accessible to us.

In this chapter we have attempted to give the reader a brief synopsis of the history of the investigation of cell cycle regulation and the status of the field up until five years ago. The last half decade, however, has seen as much progress in the field as in all the years before, and as such the balance of this book will be devoted to this recent work.

References

1 Murray AW, Hunt T. The Cell Cycle: An Introduction. New York: WH Freeman and Co., 1993.
2 Rao PN, Johnson RT. Mammalian cell fusion: studies on the regulation of DNA synthesis and mitosis. Nature 1970; 225:159–164.
3 Johnson RT, Rao PN. Mammalian cell fusion: induction of premature chromosome condensation in interphase nuclei. Nature 1970; 226:717–722.
4 Hartwell LH, Weinert TA. Checkpoints: controls that ensure the order of cell cycle events. Science 1989; 246:629–634.
5 Murray AW. Creative blocks: cell–cycle checkpoints and feedback controls. Nature 1992; 359:599–604.
6 Evans T, Rosenthal ET, Youngblom J et al. Cyclin: a protein specified by maternal mRNA in sea urchin eggs that is destroyed at each cleavage division. Cell 1983; 33:389–396.

7 Rosenthal ET, Hunt T, Ruderman JV. Selective translation of mRNA controls the pattern of protein synthesis during early development of the Surf clam, Spisula solidissima. Cell 1980; 20:487–494.
8 Swenson KI, Farrell KM, Ruderman JV. The clam embryo protein cyclin A induces entry into M phase and the resumption of meiosis in Xenopus oocytes. Cell 1986; 47:861–870.
9 Kishimoto S, Lieberman I. Synthesis of RNA and protein required for the mitosis of mammalian cells. Exp Cell Res 1964; 36:92–101.
10 Tobey RA, Petersen DF, Anderson EC et al. Life cycle analysis of mammalian cells. 3. The inhibition of division in Chinese hamster cells by puromycin and actinomycin. Biophys J 1966; 6:567–581.
11 Donnelly GM, Sisken JE. RNA and protein synthesis required for entry of cells into mitosis and during the mitotic cycle. Exp Cell Res 1967; 46:93–105.
12 Wilt FH, Sakai H, Mazia D. Old and new protein in the formation of the mitotic apparatus in cleaving sea urchin eggs. J Mol Biol 1967; 27:1–7.
13 Young CW, Hendler FJ, Karnofsky DA. Synthesis of protein for DNA replication and cleavage events in the sand dollar embryo. Exp Cell Res 1969; 58:15–26.
14 Houk MS, Epel D. Protein synthesis during hormonally induced meiotic maturation and fertilization in starfish oocytes. Dev Biol 1974; 40:298–310.
15 Phillips SG, Rattner JB. Dependence of centriole formation on protein synthesis. J Cell Biol 1976; 70:9–19.
16 Wagenaar EB, Mazia D. The effect of emetine on first cleavage division in the sea urchin, Stongylocentrotus purpuratus. In: Dirkson E, Prescott D, eds. Cell Reproduction, ICN–UCLA Symp. on Mol. Cell. Biol. New York: Academic Press, 1978:539–545.
17 Wagenaar EB. The timing of synthesis of proteins required for mitosis in the cell cycle of the sea urchin embryo. Exp Cell Res 1983; 144:393–403.
18 Westwood JT, Wagenaar EB. Chinese hamster cells can be reversibly blocked before mitosis with the protein synthesis inhibitor, emetine. J Cell Sci 1983; 59:257–268.
19 Smith LD, Ecker RE. Role of the oocyte nucleus in physiological maturation in Rana pipiens. Dev Biol 1969; 19:281–309.
20 Smith LD, Ecker RE. The interaction of steroids with Rana pipiens oocytes in the induction of maturation. Dev Biol 1971; 25:232–247.
21 Masui Y, Markert CL. Cytoplasmic control of nuclear behavior during meiotic maturation of frog oocytes. J Exp Zool 1971; 177:129–145.
22 Kishimoto T, Kanatani H. Cytoplasmic factor responsible for germinal vesicle breakdown and meiotic maturation in starfish oocyte. Nature 1976; 260:321–322.
23 Sunkara PS, Wright DA, Rao PN. Mitotic factors from mammalian cells induce germinal vesicle breakdown and chromosome condensation in amphibian oocytes. Proc Natl Acad Sci USA 1979; 76:2799–2802.
24 Nelkin B, Nichols C, Vogelstein B. Protein factor(s) from mitotic CHO cells induce meiotic maturation in Xenopus laevis ooctyes. FEBS Lett 1980; 109:233–238.
25 Kishimoto T, Kuriyama R, Kondo H et al. Generality of the action of various maturation–promoting factors. Exp Cell Res 1982; 137:121–126.
26 Weintraub H, Buscaglia M, Ferrez M et al. Mise en evidence d'une activite "MPF" chez Saccharomyces cerevisiae. C R Acad Sci (Paris) III 1982; 295:787–790.
27 Gerhart J, Wu M, Kirschner M. Cell cycle dynamics of an M-phase–specific cytoplasmic factor in Xenopus laevis oocytes and eggs. J Cell Biol 1984; 98:1247–1255.
28 Meijer L, Guerrier P. Maturation and fertilization in starfish oocytes. Int Rev Cytol 1984; 86:129–196.
29 Sorensen RA, Cyert MS, Pedersen RA. Active maturation–promoting factor is present in mature mouse oocytes. J Cell Biol 1985; 100:1637–1640.
30 Tachibana K, Yanagishima N, Kishimoto T. Preliminary characterization of maturation–promoting factor from yeast Saccharomyces cerevisiae. J Cell Sci 1987; 88:273–281.
31 Hashimoto N, Kishimoto T. Regulation of meiotic metaphase by a cytoplasmic maturation–promoting factor during mouse oocyte maturation. Dev Biol 1988; 126:242–252.

32 Wasserman WJ, Masui Y. Effects of cyclohexamide on a cytoplasmic factor initiating meiotic maturation in Xenopus oocytes. Exp Cell Res 1975; 91:381–388.
33 Newport JW, Kirschner MW. Regulation of the cell cycle during early Xenopus development. Cell 1984; 37:731–742.
34 Lohka MJ, Masui Y. Formation in vitro of sperm pronuclei and mitotic chromosomes induced by amphibian ooplasmic components. Science 1983; 220:719–721.
35 Lohka MJ, Maller JL. Induction of nuclear envelope breakdown, chromosome condensation, and spindle formation in cell–free extracts. J Cell Biol 1985; 101:518–523.
36 Wasserman WJ, Smith LD. The cyclic behavior of a cytoplasmic factor controlling nuclear membrane breakdown. J Cell Biol 1978; 78:R15–R22.
37 Lake RS, Salzman NP. Occurrence and properties of a chromatin–associated F1–histone phosphokinase in mitotic Chinese hamster cells. Biochemistry 1972; 11:4817–4826.
38 Lake RS. Further characterization of the F1–histone phosphokinase of metaphase–arrested animal cells. J Cell Biol 1973; 58:317–331.
39 Bradbury EM, Inglis RJ, Matthews HR. Control of cell division by very lysine rich histone (F1) phosphorylation. Nature 1974; 247:257–261.
40 Zeilig CE, Langan TA. Studies on the mechanism of activation of mitotic histone H1 kinase. Biochem Biophys Res Commun 1980; 95:1372–1379.
41 Woodford TA, Pardee AB. Histone H1 kinase in exponential and synchronous populations of Chinese hamster fibroblasts. J Biol Chem 1986; 261:4669–4676.
42 Labbe JC, Picard A, Karsenti E et al. An M–phase–specific protein kinase of Xenopus oocytes: partial purification and possible mechanism of its periodic activation. Dev Biol 1988; 127:157–169.
43 Langan TA, Gautier J, Lohka M et al. Mammalian growth–associated H1 histone kinase: a homolog of cdc2+/CDC28 protein kinases controlling mitotic entry in yeast and frog cells. Mol Cell Biol 1989; 9:3860–3868.
44 Lohka MJ, Hayes MK, Maller JL. Purification of maturation–promoting factor, an intracellular regulator of early mitotic events. Proc Natl Acad Sci USA 1988; 85:3009–3013.
45 Labbe JC, Lee MG, Nurse P et al. Activation at M-phase of a protein kinase encoded by a starfish homologue of the cell cycle control gene cdc2+. Nature 1988; 335:251–254.
46 Arion D, Meijer L, Brizuela L et al. cdc2 is a component of the M phase–specific histone H1 kinase: evidence for identity with MPF. Cell 1988; 55:371–378.
47 Labbe JC, Capony JP, Caput D et al. MPF from starfish oocytes at first meiotic metaphase is a heterodimer containing one molecule of cdc2 and one molecule of cyclin B. EMBO J 1989; 8:3053-3058.
48 Gautier J, Minshull J, Lohka M et al. Cyclin is a component of maturation–promoting factor from Xenopus. Cell 1990; 60:487–494.
49 Hartwell LH, Culotti J, Reid B. Genetic control of the cell–division cycle in yeast. I. Detection of mutants. Proc Natl Acad Sci USA 1970; 66:352–359.
50 Hartwell LH, Culotti J, Pringle JR et al. Genetic control of the cell division cycle in yeast. Science 1974; 183:46–51.
51 Hartwell LH. Cell division from a genetic perspective. J Cell Biol 1978; 77:627–637.
52 Pringle JR. The use of conditional lethal cell cycle mutants for temporal and functional sequence mapping of cell cycle events. J Cell Physiol 1978; 95:393–405.
53 Nurse P. Genetic control of cell size at cell division in yeast. Nature 1975; 256:547–551.
54 Nasmyth K, Nurse P. Cell division cycle mutants altered in DNA replication and mitosis in the fission yeast Schizosaccharomyces pombe. Mol Gen Genet 1981; 182:119–124.
55 Forsburg SL, Nurse P. Cell cycle regulation in the yeasts Saccharomyces cerevisiae and Schizosaccharomyces pombe. Annu Rev Cell Biol 1991; 7:227–256.
56 Reed SI. The selection of S. cerevisiae mutants defective in the start event of cell division. Genetics 1980; 95:561–577.
57 Pringle JR, Hartwell LH. The Saccharomyces cerevisiae cell cycle. In: Strathern JN, Jones EW, Broach JR, eds. The Molecular Biology of the Yeast Saccharomyces, I. Life Cycle and Inheritance. New York: Cold Spring Harbor Laboratories, 1981:97–142.

58 Nurse P, Bissett Y. Gene required in G1 for commitment to cell cycle and in G2 for control of mitosis in fission yeast. Nature 1981; 292:558–560.
59 Bartlett R, Nurse P. Yeast as a model system for understanding the control of DNA replication in eukaryotes. Bioessays 1990; 12:457–463.
60 Nasmyth KA. FAR–reaching discoveries about the regulation of START. Cell 1990; 63:1117–1120.
61 Hartwell LH, Mortimer RK, Culotti J et al. Genetic control of the cell division cycle in yeast: V. Genetic analysis of cdc mutants. Genetics 1973; 74:267–287.
62 Piggott JR, Rai R, Carter BL. A bifunctional gene product involved in two phases of the yeast cell cycle. Nature 1982; 298:391–393.
63 Reed SI, Wittenberg C. Mitotic role for the Cdc28 protein kinase of Saccharomyces cerevisiae. Proc Natl Acad Sci USA 1990; 87:5697–5701.
64 Nurse P, Thuriaux P. Regulatory genes controlling mitosis in the fission yeast Schizosaccharomyces pombe. Genetics 1980; 96:627–637.
65 Fantes PA. Isolation of cell size mutants of a fission yeast by a new selective method: characterization of mutants and implications for division control mechanisms. J Bacteriol 1981; 146:746–754.
66 Beach D, Durkacz B, Nurse P. Functionally homologous cell cycle control genes in budding and fission yeast. Nature 1982; 300:706–709.
67 Booher R, Beach D. Site–specific mutagenesis of $cdc2^{+}$, a cell cycle control gene of the fission yeast Schizosaccharomyces pombe. Mol Cell Biol 1986; 6:3523–3530.
68 Draetta G, Brizuela L, Potashkin J et al. Identification of p34 and p13, human homologs of the cell cycle regulators of fission yeast encoded by $cdc2^{+}$ and $suc1^{+}$. Cell 1987; 50:319–325.
69 Lee MG, Nurse P. Complementation used to clone a human homologue of the fission yeast cell cycle control gene cdc2. Nature 1987; 327:31–35.
70 Cisek LJ, Corden JL. Phosphorylation of RNA polymerase by the murine homologue of the cell–cycle control protein cdc2. Nature 1989; 339:679–684.
71 John PC, Sek FJ, Lee MG. A homolog of the cell cycle control protein $p34^{cdc2}$ participates in the division cycle of Chlamydomonas, and a similar protein is detectable in higher plants and remote taxa. Plant Cell 1989; 1:1185–1193.
72 Krek W, Nigg EA. Structure and developmental expression of the chicken CDC2 kinase. EMBO J 1989; 8:3071–3078.
73 Feiler HS, Jacobs TW. Cell division in higher plants: a cdc2 gene, its 34–kDa product, and histone H1 kinase activity in pea. Proc Natl Acad Sci USA 1990; 87:5397–5401.
74 Jimenez J, Alphey L, Nurse P et al. Complementation of fission yeast $cdc2^{ts}$ and $cdc25^{ts}$ mutants identifies two cell cycle genes from Drosophila: a cdc2 homologue and string. EMBO J 1990; 9:3565–3571.
75 John PC, Sek FJ, Carmichael JP et al. $p34^{cdc2}$ homologue level, cell division, phytohormone responsiveness and cell differentiation in wheat leaves. J Cell Sci 1990; 97:627–630.
76 Lehner CF, O'Farrell PH. The roles of Drosophila cyclins A and B in mitotic control. Cell 1990; 61:535–547.
77 Colasanti J, Tyers M, Sundaresan V. Isolation and characterization of cDNA clones encoding a functional $p34^{cdc2}$ homologue from Zea mays. Proc Natl Acad Sci USA 1991; 88:3377–3381.
78 Ferreira PC, Hemerly AS, Villarroel R et al. The Arabidopsis functional homolog of the $p34^{cdc2}$ protein kinase. Plant Cell 1991; 3:531–540.
79 Hirt H, Pay A, Gyorgyey J et al. Complementation of a yeast cell cycle mutant by an alfalfa cDNA encoding a protein kinase homologous to $p34^{cdc2}$. Proc Natl Acad Sci USA 1991; 88:1636–1640.
80 Hindley J, Phear GA. Sequence of the cell division gene CDC2 from Schizosaccharomyces pombe; patterns of intron splicing and homology to protein kinases. Gene 1984; 31:129–134.
81 Lorincz AT, Reed SI. Primary structure homology between the product of yeast cell cycle control gene CDC28 and vertebrate oncogenes. Nature 1984; 307:183–185.
82 Reed SI, Hadwiger JA, Lorincz AT. Protein kinase activity associated with the product of the yeast cell division cycle gene CDC28. Proc Natl Acad Sci USA 1985; 82:4055–4059.
83 Simanis V, Nurse P. The cell cycle control gene $cdc2^{+}$ of fission yeast encodes a protein kinase potentially regulated by phosphorylation. Cell 1986; 45:261–268.

84 Nurse P, Thuriaux P, Nasmyth K. Genetic control of the cell division cycle in the fission yeast Schizosaccharomyces pombe. Mol Gen Genet 1976; 146:167–178.
85 Booher R, Beach D. Interaction between cdc13+ and cdc2+ in the control of mitosis in fission yeast; dissocation of the G1 and G2 roles of the cdc2+ protein kinase. EMBO J 1987; 6:3441–3447.
86 Solomon M, Booher R, Kirschner M et al. Cyclin in fission yeast. Cell 1988; 54:738–739.
87 Goebl M, Byers B. Cyclin in fission yeast. Cell 1988; 54:739–740.
88 Hagan I, Hayles J, Nurse P. Cloning and sequencing of the cyclin–related cdc13+ gene and a cytological study of its role in fission yeast mitosis. J Cell Sci 1988; 91:587–595.
89 Ghiara JB, Richardson HE, Sugimoto K et al. A cyclin B homolog in S. cerevisiae: chronic activation of the Cdc28 protein kinase by cyclin prevents exit from mitosis. Cell 1991; 65:163–174.
90 Surana U, Robitsch H, Price C et al. The role of CDC28 and cyclins during mitosis in the budding yeast S. cerevisiae. Cell 1991; 65:145–161.
91 Hunt T. Maturation promoting factor, cyclin and the control of M-phase. Curr Opin Cell Biol 1989; 1:268–274.
92 Moreno S, Hayles J, Nurse P. Regulation of $p34^{cdc2}$ protein kinase during mitosis. Cell 1989; 58:361–372.
93 Booher RN, Alfa CE, Hyams JS et al. The fission yeast cdc2/cdc13/suc1 protein kinase: regulation of catalytic activity and nuclear localization. Cell 1989; 58:485–497.
94 Dunphy WG, Brizuela L, Beach D et al. The Xenopus cdc2 protein is a component of MPF, a cytoplasmic regulator of mitosis. Cell 1988; 54:423–431.
95 Gautier J, Norbury C, Lohka M et al. Purified maturation–promotion factor contains the product of a Xenopus homolog of the fission yeast cell cycle control gene cdc2+. Cell 1988; 54:433–439.
96 Riabowol K, Draetta G, Brizuela L et al. The cdc2 kinase is a nuclear protein that is essential for mitosis in mammalian cells. Cell 1989; 57:393–401.
97 Th'ng JP, Wright PS, Hamaguchi J et al. The FT210 cell line is a mouse G2 phase mutant with a temperature–sensitive CDC2 gene product. Cell 1990; 63:313–324.
98 Pines J, Hunter T. The differential localization of human cyclins A and B is due to a cytoplasmic retention signal in cyclin B. EMBO J 1994; 13:3772–3781.
99 Jackman M, Firth M, Pines J. Human cyclins B1 and B2 are localized to strikingly different structures: B1 to microtubules, B2 primarily to the Golgi apparatus. EMBO J 1995; 14:1646–1654.
100 Draetta G, Luca F, Westendorf J et al. Cdc2 protein kinase is complexed with both cyclin A and B: evidence for proteolytic inactivation of MPF. Cell 1989; 56:829–838.
101 Minshull J, Golsteyn R, Hill CS et al. The A- and B-type cyclin associated cdc2 kinases in Xenopus turn on and off at different times in the cell cycle. EMBO J 1990; 9:2865–2875.
102 Parker LL, Atherton–Fessler S, Lee MS et al. Cyclin promotes the tyrosine phosphorylation of $p34^{cdc2}$ in a wee1+ dependent manner. EMBO J 1991; 10:1255–1263.
103 Peeper DS, Parker LL, Ewen ME et al. A- and B-type cyclins differentially modulate substrate specificity of cyclin–cdk complexes. EMBO J 1993; 12:1947–1954.
104 Dunphy WG, Newport JW. Fission yeast p13 blocks mitotic activation and tyrosine dephosphorylation of the Xenopus cdc2 protein kinase. Cell 1989; 58:181–191.
105 Pines J, Hunter T. Human cyclin A is adenovirus E1A–associated protein p60 and behaves differently from cyclin B. Nature 1990; 346:760–763.
106 Elledge SJ, Spottswood MR. A new human p34 protein kinase, CDK2, identified by complementation of a cdc28 mutation in Saccharomyces cerevisiae, is a homolog of Xenopus Eg1. EMBO J 1991; 10:2653–2659.
107 Paris J, Le Guellec R, Couturier A et al. Cloning by differential screening of a Xenopus cDNA coding for a protein highly homologous to cdc2. Proc Natl Acad Sci USA 1991; 88:1039–1043.
108 Milarski KL, Dunphy WG, Russell P et al. Cloning and characterization of Xenopus cdc2, a component of MPF. Cold Spring Harb Symp Quant Biol 1991; 56:377–384.
109 Ninomiya–Tsuji J, Nomoto S, Yasuda H et al. Cloning of a human cDNA encoding a CDC2–related kinase by complementation of a budding yeast cdc28 mutation. Proc Natl Acad Sci USA 1991; 88:9006–9010.
110 Tsai LH, Harlow E, Meyerson M. Isolation of the human cdk2 gene that encodes the cyclin A– and adenovirus E1A–associated p33 kinase. Nature 1991; 353:174–177.

111 Fang F, Newport JW. Evidence that the G1–S and G2–M transitions are controlled by different cdc2 proteins in higher eukaryotes. Cell 1991; 66:731–742.
112 Pagano M, Pepperkok R, Verde F et al. Cyclin A is required at two points in the human cell cycle. EMBO J 1992; 11:961–971.
113 Tsai LH, Lees E, Faha B et al. The cdk2 kinase is required for the G1–to–S transition in mammalian cells. Oncogene 1993; 8:1593–1602.
114 Hadwiger JA, Wittenberg C, Richardson HE et al. A family of cyclin homologs that control the G1 phase in yeast. Proc Natl Acad Sci USA 1989; 86:6255–6259.
115 Sudbery PE, Goodey AR, Carter BL. Genes which control cell proliferation in the yeast Saccharomyces cerevisiae. Nature 1980; 288:401–404.
116 Carter BL, Sudbery PE. Small–sized mutants of Saccharomyces cerevisiae. Genetics 1980; 96:561–566.
117 Cross FR. DAF1, a mutant gene affecting size control, pheromone arrest, and cell cycle kinetics of Saccharomyces cerevisiae. Mol Cell Biol 1988; 8:4675–4684.
118 Nash R, Tokiwa G, Anand S et al. The $WHI1^{+}$ gene of Saccharomyces cerevisiae tethers cell division to cell size and is a cyclin homolog. EMBO J 1988; 7:4335–4346.
119 Richardson HE, Wittenberg C, Cross F et al. An essential G1 function for cyclin–like proteins in yeast. Cell 1989; 59:1127-1133.
120 Cross FR. Cell cycle arrest caused by CLN gene deficiency in Saccharomyces cerevisiae resembles START-I arrest and is independent of the mating–pheromone signalling pathway. Mol Cell Biol 1990; 10:6482-6490.
121 Wittenberg C, Sugimoto K, Reed SI. G1–specific cyclins of S. cerevisiae: cell cycle periodicity, regulation by mating pheromone, and association with the $p34^{CDC28}$ protein kinase. Cell 1990; 62:225–237.
122 Lew DJ, Dulic V, Reed SI. Isolation of three novel human cyclins by rescue of G1 cyclin (Cln) function in yeast. Cell 1991; 66:1197–1206.
123 Forsburg SL, Nurse P. Identification of a G1–type cyclin $puc1^{+}$ in the fission yeast Schizosaccharomyces pombe. Nature 1991; 351:245–248.
124 Koff A, Cross F, Fisher A et al. Human cyclin E, a new cyclin that interacts with two members of the CDC2 gene family. Cell 1991; 66:1217–1228.
125 Matsushime H, Roussel MF, Ashmun RA et al. Colony–stimulating factor 1 regulates novel cyclins during the G1 phase of the cell cycle. Cell 1991; 65:701–713.
126 Motokura T, Bloom T, Kim HG et al. A novel cyclin encoded by a bcl1–linked candidate oncogene. Nature 1991; 350:512-515.
127 Xiong Y, Connolly T, Futcher B et al. Human D-type cyclin. Cell 1991; 65:691–699.
128 Pardee AB. A restriction point for control of normal animal cell proliferation. Proc Natl Acad Sci USA 1974; 71:1286–1290.

Chapter 3

Cyclins and Cyclin–Dependent Kinases

A growing number of cyclins and cyclin–dependent kinases (CDKs) have been identified in a diverse group of organisms ranging from yeasts to humans. The first cyclins and CDKs, as we related in the last chapter, were discovered by virtue of their remarkable properties—cyclins A and B because they undergo continuous accumulation punctuated by periodic destruction, CDK1 because its mutations cause cell cycle arrest. Subsequently, proteins were classified as cyclins and CDKs because of their functional or structural similarities to previously known cyclins and CDKs.

A study of the origins of the D–type cyclins provides an excellent illustration of how investigators have uncovered new members of these classes of cell cycle regulators. Several groups identified members of the gene family essentially simultaneously but used widely different strategies. For example, human cyclin D was isolated because it rescued *CLN*–deficient *Saccharomyces cerevisiae* from G1 arrest, demonstrating it could functionally substitute for cyclins.[1,2] Another group identified mouse cyclin D1 in a search for genes regulated by colony–stimulating factor 1 (CSF–1) during the macrophage cell cycle.[3] Yet another isolation of cyclin D resulted from work aimed at identifying the *PRAD1* oncogene, thought to promote the formation of parathyroid tumors. Using molecular genetics, it was found that in many of these tumors the parathyroid hormone enhancer was joined to a cyclin–like gene by virtue of a chromosome translocation; this cyclin–like, oncogenic target was quickly characterized as cyclin D1.[4] Human cyclin D3 was isolated when its gene cross–hybridized to the cyclin D1 gene in a cDNA screen intended to find close relatives of cyclin D1.[5] In none of these instances was a D-type cyclin identified because it displayed unusual periodic behavior similar to that of cyclins A and B. Indeed, although cyclin D can be induced by agonists that promote proliferation, the protein does not oscillate in any predictable way during the cell cycle in cells proliferating asynchronously (i.e., cells that are not traversing the cell cycle in a coordinated manner dictated by external stimuli, e.g., serum stimulation). One can therefore argue about the semantics of using the name "cyclin." It is important to keep in mind as we proceed through this chapter that the cyclins and CDKs are classified as families of proteins not because of behavioral similarities but rather because of shared structural and functional properties. Many of the proteins to be discussed as of yet have no clear function, but they are nevertheless considered to be cyclins or CDKs solely because of sequence homologies, shared motifs and common structural elements. As such, when we describe characteristics of cyclins and CDKs without making reference to individual members of either family, we are referring to generalized structural features shared among many cyclins and CDKs without regard to any specific roles they may play in various intervals of the cell cycle.

A number of proteins with features characteristic of cyclins have been identified, primarily in yeasts and vertebrates. Our chief concern in this book is the deregulation of the cell cycle in human cells, leading to human cancers, and thus we will devote our attention almost exclusively to those cyclins discovered in mammalian systems. If we were to arbitrarily divide the known members of the mammalian cyclin family (cyclins A through I) into subfamilies, we would be hard pressed to find some clear and indisputable basis by which to do so. The roles of half of the cyclins are so vaguely understood that the proteins defy any easy classification. The others we can tentatively separate into three groups. The mitotic cyclins, consisting of cyclins A and B, are distinguishable because their levels slowly accumulate during interphase and reach their peaks in late G2 phase or M phase. Moreover, the manner in which they are destroyed is unique to the two cyclins. The G1 cyclins, consisting of cyclins D1, D2, D3 and E, reach their maximum levels in mid–to–late G1 phase. Moreover, they are short–lived proteins in comparison to the mitotic cyclins. Rather than amassing slowly during the extent of the cell cycle, they are seen to be induced in rapid response to environmental signals originating outside of the cell and are just as quickly degraded. Cyclin H resides in a group of its own; it appears to be expressed throughout the cell cycle and has the intriguing behavior of influencing other cyclins. These groupings are far from ideal; as will become clear in the following chapters, cyclin A shares functional characteristics with both cyclin B and the G1 cyclins, and cyclins H and C are similar in that they both may play roles in transcriptional regulation. Nevertheless, our classification will serve as a good starting point from which to launch our study of cell cycle regulators.

The presently known CDKs (CDK1 through CDK8) suffer from the same problem as cyclins in that many are still poorly understood. The ones that have been fairly well characterized are not easily distinguished from one another judging solely by their intrinsic attributes. Instead we will find that these CDKs are better described in terms of the cyclin partners with which they associate in vivo. Following this prescription we can consider CDK1 to be a mitotic CDK because it binds cyclin B; CDK2, CDK4 and CDK6 can be thought of as G1/S CDKs by virtue of their binding with cyclins A and E or D.

Our intent in this chapter is merely to introduce the cast of characters which will dominate the discussion in the remainder of the book. In the text that follows we will describe the properties of these proteins, after which we will examine the general mechanism by which a cyclin and a CDK interact to form a complex. We will then be sufficiently acquainted with the actors to plunge into a detailed treatment of cyclin–CDK regulation in the subsequent chapters.

Mitotic Cyclins

A wide variety of eukaryotes express homologues of the two cyclins (A and B) originally discovered in *Lytechinus pictus* and *Spisula solidissima*; these proteins are collectively called A-type cyclins and B-type cyclins. The earliest work found that cyclins A and B oscillate during the cell cycle, reaching maximal levels at slightly different times during early mitosis before undergoing rapid degradation.[6] Moreover, cyclins A and B are equally capable of inducing M phase when injected into *Xenopus laevis* oocytes,[7-10] earning them the label of "mitotic cyclins." Notably, these studies were carried out in meiotic or embryonic cells, which have abbreviated cell cycles consisting of alternating S and M phases with no intervening gap phases. In somatic cells the expression of cyclin A commences before that of cyclin B, such that cyclin A is largely synthesized during S phase, cyclin B in G2 phase. The lack of simultaneity in expression suggests that the two cyclins are functionally distinct.

A wealth of experimental evidence substantiates this supposition. Genetic studies in *Drosophila* fruit flies indicate that the two cyclins play different roles in development, for while they are both expressed in embryos, a deficiency in cyclin A will cause cell proliferation to cease at the 15th cell cycle.[11,12] The presence of cyclin B alone does not avert the arrested development. Cyclin A is localized in the nucleus throughout the cell cycle;[13-18] in contrast, cyclin B appears to reside in the cytoplasm until the onset of mitosis, when it is translocated to the nucleus.[14,17,19-21] A-type cyclin–CDK complexes exhibit increased enzymatic activity earlier in the cell cycle than B–type cyclin–CDK complexes, and the two types of complexes have different substrate specificities.[22,23] Moreover, cyclin A is required in both S and G2 phases for the human cell cycle to proceed; the only mandatory role of cyclin B occurs at the end of G2.[13,15,16]

In attempting to make sense of these data distinguishing the two cyclins, the most striking observation is that across the various species, the cyclins that serve as the regulatory subunits of maturation–promoting factor, MPF, are all of the B–type variety.[24] Cyclin A readily binds to the CDK (CDK1) component of MPF in vivo and in vitro,[22,25] but cannot substitute for cyclin B in originating MPF activity. Unlike cyclin B, cyclin A binds to another CDK (CDK2) in vivo, conferring upon it functions distinct from those of cyclin B.[26,27] CDK2 activity is not limited to G2 phase and mitosis, as is CDK1 activity, but is seen in S phase as well,[28-30] implicating cyclin A as an "S phase cyclin."

While it seems likely that cyclin A has an essential function as a mitotic cyclin, the evidence implies that cyclin A is important in regulating poorly understood checkpoints associated with the replication of DNA. As we proceed through later chapters and learn about the relationships between the cyclins and other cell cycle regulators, it will become clear that cyclin A is a multifunctional protein. It will be helpful for the reader to think of cyclin B as the "true" mitotic cyclin, whereas cyclin A splits its duties between serving as a partner of cyclin B in late G2 phase and fulfilling separate, critical roles in the S and early G2 phases.

The most distinctive feature of cyclins A and B when compared to all other cyclins is their means of destruction. Both undergo rapid degradation soon after their levels peak near the G2/M phase transition, thereby resetting the cell cycle clock at the start of the interphase following the mitosis. Indeed, it has become clear that this destruction *must* occur if the cell is to complete mitosis, for mutants of both cyclins that are rendered invulnerable to degradation cause a cell cycle arrest in M phase.[20,31,32] This requirement ensures that no active MPF survives past the end of mitosis and immediately precipitates a second (and disastrously premature) mitosis.

All known homologues of cyclins A and B possess an N-terminal sequence called the "destruction box" because it is both necessary and sufficient for the protein which contains it to be targeted for degradation.[33] The mechanism through which the degradation occurs is ubiquitin–dependent proteolysis.[33,34] The destruction box, generally found to be adjacent to a lysine–rich sequence, acts as a recognition marker that directs special enzymes to conjugate a ubiquitin molecule to a lysine residue on the cyclin. More ubiquitins are linked to the first one, and eventually the cyclin comes into contact with a proteasome, a specialized construct that cleaves polyubiquitinated proteins into small peptides. Exactly what induces the ubiquitination of the mitotic cyclins in M phase (as opposed to any other stage of the cell cycle) is presently unclear, although one may reasonably suppose that the process is linked to MPF activity. Recent work has identified the anaphase–promoting complex (APC) or cyclosome, a large ubiquitin ligase complex, as the major component of the proteolytic system that degrades cyclins prior to the initiation of anaphase and exit from mitosis. This complex is the subject of intense research at present, and the subunit composition of the APC is just now being described.[35-37] While clearly of great importance in mitotic control, a biochemical understanding of the APC, its regulation and activities await much further analysis.

Although the ubiquitin–mediated proteolysis pathway appears to be important not only for the destruction of cyclins A and B but also for other components of the cell cycle machinery, it does not appear at present to bear directly on the involvement of cell cycle regulators in cancer and thus lies beyond the scope of this book. It should nevertheless be noted that the past few years have witnessed significant progress in elucidating the ubiquitin pathway as well as the role of cyclin degradation in the regulation of mitotic events, and we would be remiss not to direct interested readers to several recent review articles on these subjects.[35,38-40]

G1 Cyclins

As we related in the introduction to this chapter, the D–type cyclins were simultaneously identified by a number of independent methods.[1-5,41] The three members of this cyclin subfamily are designated cyclins D1, D2 and D3. Whether these closely related proteins function redundantly during the cell cycle or have separate roles is unclear. The murine versions of the three proteins are significantly homologous to the human versions, to the extent that each mouse cyclin is less closely related to the other mouse cyclins than to its human cognate.[41,42] This high degree of conservation suggests that the proteins have important, distinct functions. Yet the D–type cyclins are differentially expressed in various combinations in various cell lines, such that no single member seems to be essential in every cell type; the implication is that the proteins have overlapping roles and that a subset of the family can substitute for a complete panel of family members.[3,5,43-47]

Despite their sequence homologies, the D–type cyclins are a study of contrasts with the mitotic cyclins with respect to their behavioral properties. They are very labile, with a half–life of less than thirty minutes, and they are synthesized not in a cell–cycle–dependent manner but only during the exposure of the cell to growth–promoting mitogens.[3,43,46,48-50] Indeed, cyclin D–CDK activity quickly vanishes upon the withdrawal of mitogens. The D–type cyclins appear to be more responsive and closely linked to growth stimuli rather than the internal mechanics of the cell cycle.

Nevertheless, it is clear that cyclin D is a pivotal actor in cell cycle regulation. Overexpression of D–type cyclins in rat fibroblasts abbreviates the time the cells spend in G1 phase, although the presence of mitogens is still required for the G1/S phase transition to be surpassed.[51-53] Moreover, the insertion of either antisense oligonucleotides to the cyclin D1 gene or antibodies to cyclin D1 protein into fibroblasts during mid–G1 phase blocks the cells' ability to progress into S phase, although the effect disappears if the insertion takes place when the cells are close to the G1/S boundary.[51,54] This evidence pinpoints a crucial role for the D-type cyclins during mid–to–late G1 phase. Also falling into this interval is the restriction (R) point, arguably the most important checkpoint in the entire mammalian cell cycle; after a cell has proceeded past the R point, it cannot exit the cell cycle to enter quiescent or senescent states but is restricted to completing a round of DNA replication and the subsequent mitosis. Since the two events are roughly contemporaneous, it is not unreasonable to assert that the activation of cyclin D–CDK complexes initiates processes that enable the cell to pass the R point.

CDK4 and CDK6 were originally characterized as catalytic subunits associating specifically with cyclin D, and subsequent investigation has demonstrated that although individual D–type cyclins can complex with CDK2 and CDK5 in certain circumstances, CDK4 and CDK6 are the preferred binding partners of cyclin D in vivo, and vice versa.[48,55-61] This pair of CDKs is distinct from others because of their ability to preferentially phosphorylate the retinoblastoma susceptibility gene product (pRb) and their disinterest in histone H1. The significance of this will be discussed in chapter 6.

Cyclin E was independently discovered by two groups, in both cases by virtue of its ability to rescue *CLN*–deficient *S. cerevisiae* from G1 arrest.[1,62] Whereas cyclin D levels are determined by the presence or absence of mitogens, cyclin E is expressed in a cell–cycle–dependent manner similar to cyclins A and B. Unlike the latter cyclins, cyclin E largely accumulates during G1 phase, reaches its maximal level in early S phase and undergoes degradation shortly thereafter.[1,63,64] Notably, cyclin E is usually observed to peak after the D-type cyclins during G1 phase.

Cyclin E's primary binding partner appears to be CDK2 rather than CDK4 or CDK6. Cyclin E–CDK2 activity and abundance both reach their maximal levels around the G1/S phase transition.[62-64] Importantly, CDK2 is required for mammalian cells to progress into S phase;[65] although CDK2 is promiscuous in that it binds both cyclins E and A, because cyclin A does not reach a comparable level to that of cyclin E until S phase it seems probable that cyclin E–CDK2 executes a crucial function prior to and linked to the G1/S phase transition. By analogy with cyclin B's role as a regulatory component of MPF, cyclin E might be a subunit of cyclin–CDK complexes acting as a "synthesis–promoting factor," or SPF. In this scenario, cyclin E would serve as the primary regulator of the G1/S phase transition. Alternatively, cyclin E-CDK2 complexes might function in the initiation of DNA synthesis (with other cyclin–CDK complexes performing the role of SPF). Despite the present uncertainty as to the precise functions of cyclins D and E, studies have established that cyclins D and E have distinct, separable roles in G1 phase, both of which must be fulfilled in order for the cell cycle to advance.[66,67]

The D-type cyclins and cyclin E appear to be degraded by a pathway distinct from cyclins A and B, for they do not contain the canonical destruction box that facilitates ubiquitin–mediated proteolysis. They instead share a different motif: each contains a C-terminal region rich in proline (P), glutamate (E), serine (S) and threonine (T) residues. Commonly known as the PEST sequence, this region is thought to destabilize any protein which contains it, such that the protein has an extremely short half life (<2 hours).[68] Moreover, unlike the destruction box, the PEST sequence renders the protein vulnerable to degradation throughout the cell cycle; in order to maintain a PEST protein at a constant level, there must be continuous synthesis to balance the continuous loss of the protein. How PEST targets a protein for rapid destruction is unclear, although ubiquitin is suspected to be an accomplice. Not surprisingly, the removal of PEST from a cyclin translates into an elevated cyclin level in the cell.[69,70] Recent work shows that the PEST domain of cyclin E contains a regulatory CDK2 phosphorylation site that leads to dissociation of cyclin E from CDK2, followed by ubiquitination and degradation of the "free" cyclin in an apparent post–translational autoregulation of cyclin E-CDK2 activity.[71,72]

The presence of the PEST sequence in cyclins D and E is consistent with their transient presence during limited intervals of the cell cycle. So many of the other mammalian cyclins carry the PEST sequence that it is commonly considered a generalized feature of the cyclin family; it is curious that the original, prototypical cyclins, A and B, should be exceptions to the rule.

Miscellaneous Cyclins and CDKs

Of the remaining cyclins and CDKs, cyclin H and CDK7 are the best understood. The *CDK7* gene, originally known as *MO15*, was first cloned as a gene encoding a CDK1–related protein kinase involved in the negative regulation of meiotic maturation of *Xenopus laevis* oocytes.[73] It was soon discovered to be the catalytic component of an enzyme activity that phosphorylated *X. laevis* cyclin B–CDK1.[74-76] Several groups rapidly identified a human homologue of *MO15*.[77-80] The encoded kinase was found to bind another protein whose presence was required for full enzymatic activity; this second protein displayed structural similarities to other cyclins.[81,82] Accordingly, the *MO15* product was accepted as a bona fide CDK, and its partner was labeled cyclin H. This cyclin–CDK pair is unique in that its primary catalytic targets appear to be other cyclin–CDK complexes. Interestingly, recent evidence suggests that cyclin H–CDK7 also participates in DNA transcription.[83-85] A further discussion of the relationship of cyclin H–CDK7 to the cell cycle (and perhaps other cellular processes) will be given in the next chapter.

The gene encoding cyclin C was identified as both a human gene and a *Drosophila* gene whose expression could overcome the G1 phase arrest of *CLN*–deficient yeast.[1,86,87] The mRNA expression level of the human gene in HeLa cells was found to moderately increase during G1 phase and then slowly decrease throughout the remainder of the cell cycle, suggesting that cyclin C might have a G1 role like cyclins D and E.[1] No cell cycle dependence was observed for the *Drosophila* version of the gene.[86,87] Neither homologue contains a cyclin destruction box, whereas the human gene possesses the PEST motif near the C-terminal end of the molecule. Recently, a new CDK (CDK8) was isolated and found to preferentially bind to cyclin C both in vitro and in vivo.[88] Because of similarities of cyclin C-CDK8 to a yeast cyclin–CDK pair involved in the regulation of RNA polymerase II, investigators have hypothesized that cyclin C might be associated with the mammalian transcription apparatus.[88,89]

Cyclin G was isolated serendipitously in a search for relatives of *src* kinase in the rat system.[90] Upon sequencing the gene, the protein was noted to have significant homology to cyclin A as well as similarities to other members of the cyclin family. More recently two human homologues of cyclin G, designated G1 and G2, have been identified.[91,92] Cyclin G1 mRNA expression appears to rise during G1 phase and, depending on the system used, either remains elevated or falls off during the remainder of the cell cycle; in contrast, cyclin G2 expression oscillates during the cell cycle, peaking in S phase. Cyclin G1 has neither a destruction box nor a PEST sequence, leaving no clear provision for its degradation, whereas cyclin G2 possesses a PEST–rich region.[90,92] Yet another distinction between the homologues lies in their tissue specificities. Cyclin G1 mRNA expression is greatest in skeletal and cardiac muscle, ovary, and kidney, whereas the highest expression of cyclin G2 occurs in cerebellum, thymus, spleen, prostate and kidney.[92]

Interestingly, cyclin G was independently discovered in a screen for mouse genes induced by the p53 tumor suppressor protein/transcription factor (see chapters 5 and 7 for a detailed discussion of this protein), and subsequent investigation established that the cyclin is a probable transcriptional target of p53.[93,94] One possible interpretation is that cyclin G is a downstream participant in p53–mediated growth suppression, marking the cyclin as a negative regulator of proliferation, but it is also conceivable that cyclin G acts antagonistically to p53 in a negative feedback loop. Other studies have shown that cyclin G is overexpressed in human osteosarcoma cells and that introduction of antisense cyclin G1 inhibits the growth of these tumor cells, raising the suspicion that cyclin G may have oncogenic properties.[91,95] No CDK partner for cyclin G has yet been identified. Any definite roles for cyclin G1 or G2 in cell cycle regulation and the distinctions between their roles remain to be unmasked.

Cyclin F was identified as a protein that complements a temperature–sensitive mutant of the *CDC4* cell division cycle gene (which causes a cell cycle arrest prior to the initiation of DNA replication) in *S. cerevisiae*; the cyclin was also serendipitously found in an independent (and unrelated) search for genes responsible for polycystic kidney disease.[96,97] Cyclin F is the largest protein of the cyclin family and shows the closest similarities to cyclins A and B, although it has the PEST motif rather than the destruction box. It appears to be expressed ubiquitously in human tissues and predominantly resides in the nucleus.[96] The cyclin F level fluctuates in a manner very similar to cyclin A: accumulating during interphase, peaking in G2 phase, and disappearing in mitosis; cyclin F is therefore thought to function during S or G2 phase. This belief is buttressed by the observation that the introduction of either overexpressed cyclin F or stabilized cyclin F (whose PEST sequence has been removed) cause cells to aggregate in G2 phase.[96] Because no CDKs have yet been observed to partner with cyclin F, an elucidation of the cyclin's precise role awaits future investigation.

The cyclin I gene was isolated from a cDNA library derived from human brain.[98] Found to be most pronounced in differentiated tissues such as skeletal muscle, heart, and brain, cyclin I expression was shown to be virtually independent of the cell cycle, giving no clue as to its function in the cell. The protein most closely resembles cyclins G and E and contains the typical C–terminal PEST sequence. No partner CDK has yet been identified.

Human CDK3 was discovered by its ability to rescue *S. cerevisiae* with CDK1 mutations, suggesting an overlap of function between the two proteins.[99] CDK3 is structurally very closely related to CDK1 and CDK2, and the expression of a dominant–negative CDK3 mutant in several human cell lines causes a G1 cell cycle arrest.[99,100] These observations indicate that the kinase plays an essential role in regulating the progression of the cell from G1 phase to S phase, perhaps through an interaction with transcription factors involved in this transition.[101] Yet the endogenous expression of CDK3 in human cells is minimal, and no cyclin that binds to the kinase in vivo has been described.[100]

CDK5 was identified by virtue of its structural homology to CDK1.[99] CDK5 activity is almost exclusively detected in differentiated neurons, suggesting a role for the protein in neuronal development.[102] Indeed, mice lacking CDK5 display profoundly altered brain development.[103] Cyclins D and E have been reported to form complexes with CDK5, but no kinase activity propagated by these complexes has been demonstrated in vivo.[55,104,105] Instead the regulatory partner of CDK5 appears to be a 35 kDa protein (p35) expressed specifically in brain tissue or, in some circumstances, a 25 kDa truncated form (p25) of p35.[106-110] p35 activation of CDK5 is required for neurite outgrowth in culture and, as with CDK5, the absence of p35 in mice leads to greatly altered brain architecture apparently due to defects in neural migration.[111,112] Because p35 shares minimal sequence homology to the mammalian cyclins and is expressed in a tissue–specific rather than cell–cycle–dependent manner, it is thought to represent a new type of regulatory subunit for CDK activity. Strictly speaking, CDK5 may not be a bona fide "cyclin–dependent" kinase, but the potentially novel nature of p35–dependent activation of CDK5 awaits confirmation by structural comparison of p35 and cyclins.

The Physical Interaction of Cyclins and CDKs

All members of the cyclin family share a homologous region roughly 100 amino acids in size called the "cyclin box" (Fig. 3.1);[69] possession of this region, in fact, is enough for a protein to be named a cyclin regardless of its other properties. CDKs are even more highly homologous than cyclins, sharing a conserved catalytic core as well as a "PSTAIRE" motif, named for a short amino acid sequence exactly conserved in CDK1, CDK2, and CDK3 and closely conserved in the other CDKs (Fig. 3.2).[113] On the basis of homology alone, it should not be surprising if the cyclin box on the one hand and the PSTAIRE motif on the other hand were involved in the interaction of a cyclin and a CDK. Indeed, a body of indirect evidence drawing primarily on mutational analysis has established that amino acids contained within the cyclin box are required for cyclin binding and activation of the CDK,[114,115] and that an intact PSTAIRE motif is needed to preserve the CDK's interaction with the cyclin.[116,117]

The most detailed analysis of the cyclin–CDK connection is provided by recent crystallization studies. The three–dimensional structure of a monomeric CDK2 molecule reveals that the amino acids surrounding and including the PSTAIRE motif form a helical structure proximal to the kinase's active site, oriented in such a way as to impede ATP binding.[118] Moreover, from the C–terminal portion of the molecule there extends a structure called the T–loop that blocks access to the active site, preventing the kinase from binding a protein substrate.[118] When cyclin A comes into contact with the CDK2 molecule, both the PSTAIRE helix and the T–loop undergo significant conformational changes.[119] The cyclin box of cyclin A molds around the PSTAIRE helix, rotating the helix and aligning active site residues into positions amenable to the binding of ATP. Figures 3.1 and 3.2 show the critical residues on either molecule that enable this interaction. The T–loop shifts and no longer obstructs the entrance to the catalytic site, allowing the substrate to come into contact with the ATP. In simply forming a

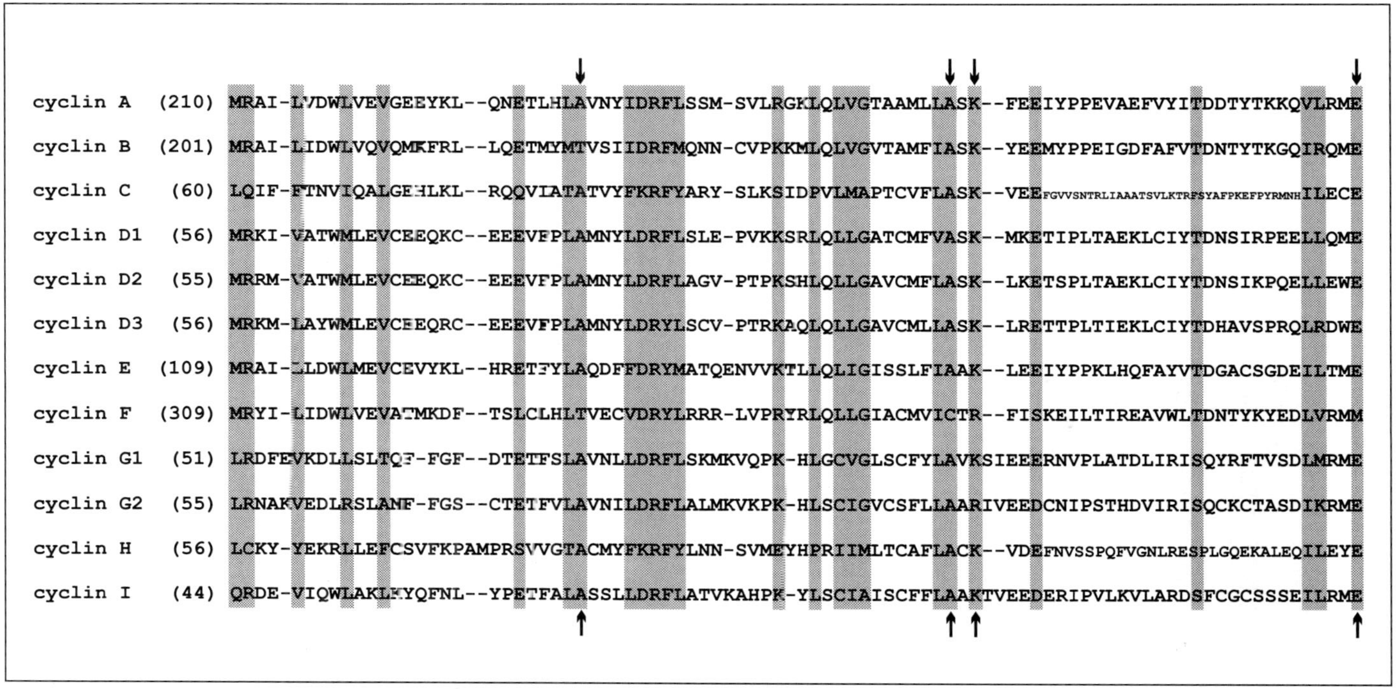

Fig. 3.1. Comparison of aligned amino acid sequences in the "cyclin box" regions of the various human cyclins. The shaded bars indicate residues that are identical or conservatively substituted (K/R, E/D, N/Q, T/S, F/Y/W, M/L/V/I) in at least nine of the twelve cyclins portrayed in the figure. The arrows indicate the residues that are thought to be most crucial for the binding of cyclins to CDKs, based on evidence from crystallization studies. All amino acid sequences are derived from the GenBank Sequence Database maintained by the National Center for Biotechnology Information, US National Library of Medicine.

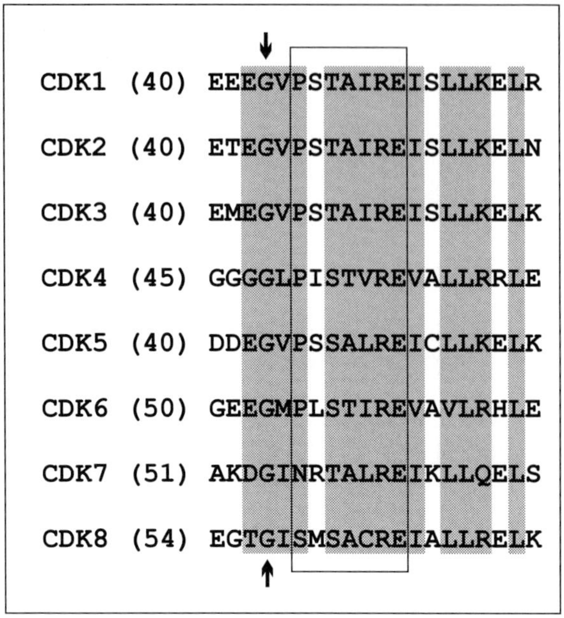

Fig. 3.2. Comparison of aligned amino acid sequences surrounding the "PSTAIRE" motifs of the various human CDKs. The shaded bars indicate residues that are identical or conservatively substituted (K/R, E/D, N/Q, T/S, F/Y/W, M/L/V/I) in at least six of the eight CDKs portrayed in the figure. The box marks off the residues composing the formal PSTAIRE motifs. The arrows indicate the residue that is thought to be most crucial for the binding of CDKs to cyclins, based on evidence from crystallization studies. All amino acid sequences are derived from the GenBank Sequence Database maintained by the National Center for Biotechnology Information, US National Library of Medicine.

complex with cyclin A, CDK2 is converted from an inactive to a potentially active kinase—potentially, because the complete story is somewhat more complicated, but we will not investigate the matter further until the next chapter.

Presumably, all cyclin–CDK complex formation proceeds in a fashion similar to cyclin A and CDK2. The portions of the molecules that are not directly involved in the binding of the molecules determine the specificity of the interaction, such that only the appropriate cyclin–CDK combinations result.

Cyclin–CDK Complexes in the Cell Cycle

Let us recap the "facts" of cyclin–CDK involvement during the cell cycle as investigators understand them at the present time (while setting aside the find-

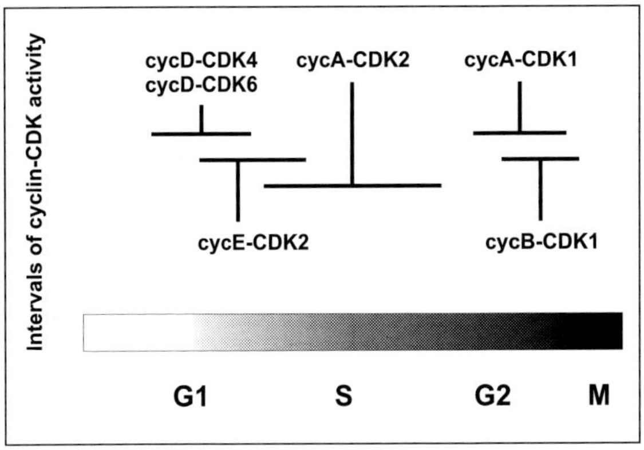

Fig. 3.3. Diagram depicting the intervals of the cell cycle during which various cyclin–CDK complexes are thought to be active. Cyclin–CDK complexes containing cyclins D and E are functional during G1 phase, cyclin A–CDK2 complexes display activity beginning at the G1/S phase transition and extending into G2 phase, and both cyclin A–CDK1 and cyclin B–CDK1 become active in late G2 phase and remain so until the destruction of the cyclins in early M phase. The boundaries of the intervals are imprecise and differ among cell types.

ings that are less firmly established). In a typical repetition of the cell cycle, cyclin D–CDK4 and cyclin D–CDK6 are the first complexes to become active, appearing during mid–to–late G1 phase. They are followed by cyclin E-CDK2 complexes in late G1 phase, which are present until after the cell has traversed the G1/S phase transition. S phase witnesses the formation of active cyclin A–CDK2 complexes; after the cell has entered G2 phase, cyclin A trades its CDK2 partner for CDK1, at which point it assumes its mitotic function. The G2/M phase transition heralds the appearance of cyclin B–CDK1, or MPF, which precipitates the events of mitosis. By the end of mitosis both cyclins A and B have been destroyed, and the cell begins G1 phase anew with a clean slate (Fig. 3.3). While much of the detail of cyclin–CDK temporal regulation remains to be worked out—e.g., exactly when and how cyclin D–CDK4 and –CDK6 complexes are inactivated after G1 phase—the basic cascade of CDK activities provides an elegant mechanism to order the cell cycle.

Having been familiarized with the properties of the various cyclins and their CDK partners, we are prepared to dissect the varied roles of cyclin–CDK complexes in cell cycle regulation. We will first examine the mechanisms by which the cell regulates cyclin–CDK activity and then discuss the downstream effectors and upstream affectors of the kinase complexes. Piece by piece we will reconstruct the cell cycle machinery and situate it in its proper context with respect to the cell's internal and external environments.

References

1 Lew DJ, Dulic V, Reed SI. Isolation of three novel human cyclins by rescue of G1 cyclin (Cln) function in yeast. Cell 1991; 66:1197–1206.

2 Xiong Y, Connolly T, Futcher B et al. Human D–type cyclin. Cell 1991; 65:691–699.

3 Matsushime H, Roussel MF, Ashmun RA et al. Colony–stimulating factor 1 regulates novel cyclins during the G1 phase of the cell cycle. Cell 1991; 65:701–713.

4 Motokura T, Bloom T, Kim HG et al. A novel cyclin encoded by a bcl1–linked candidate oncogene. Nature 1991; 350:512–515.

5 Motokura T, Keyomarsi K, Kronenberg HM et al. Cloning and characterization of human cyclin D3, a cDNA closely related in sequence to the PRAD1/cyclin D1 proto–oncogene. J Biol Chem 1992; 267:20412–20415.

6 Evans T, Rosenthal ET, Youngblom J et al. Cyclin: a protein specified by maternal mRNA in sea urchin eggs that is destroyed at each cleavage division. Cell 1983; 33:389–396.

7 Swenson KI, Farrell KM, Ruderman JV. The clam embryo protein cyclin A induces entry into M phase and the resumption of meiosis in Xenopus oocytes. Cell 1986; 47:861–870.

8 Pines J, Hunt T. Molecular cloning and characterization of the mRNA for cyclin from sea urchin eggs. EMBO J 1987; 6:2987–2995.

9 Westendorf JM, Swenson KI, Ruderman JV. The role of cyclin B in meiosis I. J Cell Biol 1989; 108:1431–1444.

10 Roy LM, Swenson KI, Walker DH et al. Activation of $p34^{cdc2}$ kinase by cyclin A. J Cell Biol 1991; 113:507–514.

11 Lehner CF, O'Farrell PH. Expression and function of Drosophila cyclin A during embryonic cell cycle progression. Cell 1989; 56:957–968.

12 Lehner CF, O'Farrell PH. The roles of Drosophila cyclins A and B in mitotic control. Cell 1990; 61:535–547.

13 Girard F, Strausfeld U, Fernandez A et al. Cyclin A is required for the onset of DNA replication in mammalian fibroblasts. Cell 1991; 67:1169–1179.

14 Pines J, Hunter T. Human cyclins A and B1 are differentially located in the cell and undergo cell cycle–dependent nuclear transport. J Cell Biol 1991; 115:1–17.

15 Pagano M, Pepperkok R, Verde F et al. Cyclin A is required at two points in the human cell cycle. EMBO J 1992; 11:961–971.

16 Zindy F, Lamas E, Chenivesse X et al. Cyclin A is required in S phase in normal epithelial cells. Biochem Biophys Res Commun 1992; 182:1144–1154.

17 Pines J, Hunter T. The differential localization of human cyclins A and B is due to a cytoplasmic retention signal in cyclin B. EMBO J 1994; 13:3772–3781.

18 Cardoso MC, Leonhardt H, Nadal–Ginard B. Reversal of terminal differentiation and control of DNA replication: cyclin A and Cdk2 specifically localize at subnuclear sites of DNA replication. Cell 1993; 74:979–992.

19 Bailly E, Pines J, Hunter T et al. Cytoplasmic accumulation of cyclin B1 in human cells: association with a detergent–resistant compartment and with the centrosome. J Cell Sci 1992; 101:529–545.

20 Gallant P, Nigg EA. Cyclin B2 undergoes cell cycle–dependent nuclear translocation and, when expressed as a non–destructible mutant, causes mitotic arrest in HeLa cells. J Cell Biol 1992; 117:213–224.

21 Ookata K, Hisanaga S, Okano T et al. Relocation and distinct subcellular localization of $p34^{cdc2}$–cyclin B complex at meiosis reinitiation in starfish oocytes. EMBO J 1992; 11:1763–1772.

22 Minshull J, Golsteyn R, Hill CS et al. The A- and B-type cyclin associated cdc2 kinases in Xenopus turn on and off at different times in the cell cycle. EMBO J 1990; 9:2865–2875.

23 Peeper DS, Parker LL, Ewen ME et al. A- and B-type cyclins differentially modulate substrate specificity of cyclin–cdk complexes. EMBO J 1993; 12:1947–1954.

24 Murray AW. Creative blocks: cell–cycle checkpoints and feedback controls. Nature 1992; 359:599–604.

25 Draetta G, Luca F, Westendorf J et al. Cdc2 protein kinase is complexed with both cyclin A and B: evidence for proteolytic inactivation of MPF. Cell 1989; 56:829–838.

26 Tsai LH, Harlow E, Meyerson M. Isolation of the human cdk2 gene that encodes the cyclin A– and adenovirus E1A–associated p33 kinase. Nature 1991; 353:174–177.

27 Elledge SJ, Richman R, Hall FL et al. CDK2 encodes a 33–kDa cyclin A-associated protein kinase and is expressed before CDC2 in the cell cycle. Proc Natl Acad Sci USA 1992; 89:2907–2911.

28 Draetta G, Beach D. Activation of cdc2 protein kinase during mitosis in human cells: cell cycle–dependent phosphorylation and subunit rearrangement. Cell 1988; 54:17–26.

29 Elledge SJ, Spottswood MR. A new human p34 protein kinase, CDK2, identified by complementation of a cdc28 mutation in Saccharomyces cerevisiae, is a homolog of Xenopus Eg1. EMBO J 1991; 10:2653–2659.

30 Fang F, Newport JW. Evidence that the G1–S and G2–M transitions are controlled by different cdc2 proteins in higher eukaryotes. Cell 1991; 66:731–742.

31 Murray AW, Solomon MJ, Kirschner MW. The role of cyclin synthesis and degradation in the control of maturation promoting factor activity. Nature 1989; 339:280–286.

32 Luca FC, Shibuya EK, Dohrmann CE et al. Both cyclin A delta 60 and B delta 97 are stable and arrest cells in M–phase, but only cyclin B delta 97 turns on cyclin destruction. EMBO J 1991; 10:4311–4320.

33 Glotzer M, Murray AW, Kirschner MW. Cyclin is degraded by the ubiquitin pathway. Nature 1991; 349:132–138.

34 Hershko A, Ganoth D, Pehrson J et al. Methylated ubiquitin inhibits cyclin degradation in clam embryo extracts. J Biol Chem 1991; 266:16376–16379.

35 Deshaies RJ. The self–destructive personality of a cell cycle in transition. Curr Opin Cell Biol 1995; 7:781–789.

36 Peters JM, King RW, Hoog C et al. Identification of BIME as a subunit of the anaphase–promoting complex. Science 1996; 274:1199–1201.

37 Zachariae W, Shin TH, Galova M et al. Identification of subunits of anaphase–promoting complex of Saccharomyces cerevisiae. Science 1996; 274:1201–1204.

38 King RW, Jackson PK, Kirschner MW. Mitosis in transition. Cell 1994; 79:563–571.

39 Hochstrasser M. Ubiquitin, proteasomes, and the regulation of intracellular protein degradation. Curr Opin Cell Biol 1995; 7:215–223.

40 Murray A. Cyclin ubiquitination: the destructive end of mitosis. Cell 1995; 81:149–152.

41 Xiong Y, Menninger J, Beach D et al. Molecular cloning and chromosomal mapping of CCND genes encoding human D-type cyclins. Genomics 1992; 13:575–584.

42 Inaba T, Matsushime H, Valentine M et al. Genomic organization, chromosomal localization, and independent expression of human cyclin D genes. Genomics 1992; 13:565–574.

43 Matsushime H, Roussel MF, Sherr CJ. Novel mammalian cyclins (CYL genes) expressed during G1. Cold Spring Harb Symp Quant Biol 1991; 56:69–74.

44 Cocks BG, Vairo G, Bodrug SE et al. Suppression of growth factor–induced CYL1 cyclin gene expression by antiproliferative agents. J Biol Chem 1992; 267:12307–12310.

45 Kiyokawa H, Busquets X, Powell CT et al. Cloning of a D–type cyclin from murine erythroleukemia cells. Proc Natl Acad Sci USA 1992; 89:2444–2447.

46 Won KA, Xiong Y, Beach D et al. Growth–regulated expression of D–type cyclin genes in human diploid fibroblasts. Proc Natl Acad Sci USA 1992; 89:9910–9914.

47 Ajchenbaum F, Ando K, DeCaprio JA et al. Independent regulation of human D–type cyclin gene expression during G1 phase in primary human T lymphocytes. J Biol Chem 1993; 268:4113–4119.

48 Matsushime H, Ewen ME, Strom DK et al. Identification and properties of an atypical catalytic subunit ($p34^{PSK-J3}$/cdk4) for mammalian D type G1 cyclins. Cell 1992; 71:323–334.

49 Winston JT, Pledger WJ. Growth factor regulation of cyclin D1 mRNA expression through protein synthesis–dependent and –independent mechanisms. Mol Biol Cell 1993; 4:1133–1144.

50 Bates S, Parry D, Bonetta L et al. Absence of cyclin D/cdk complexes in cells lacking functional retinoblastoma protein. Oncogene 1994; 9:1633–1640.

51 Quelle DE, Ashmun RA, Shurtleff SA et al. Overexpression of mouse D-type cyclins accelerates G1 phase in rodent fibroblasts. Genes Dev 1993; 7:1559–1571.

52 Jiang W, Kahn SM, Zhou P et al. Overexpression of cyclin D1 in rat fibroblasts causes abnormalities in growth control, cell cycle progression and gene expression. Oncogene 1993; 8:3447–3457.
53 Resnitzky D, Gossen M, Bujard H et al. Acceleration of the G1/S phase transition by expression of cyclins D1 and E with an inducible system. Mol Cell Biol 1994; 14:1669–1679.
54 Baldin V, Lukas J, Marcote MJ et al. Cyclin D1 is a nuclear protein required for cell cycle progression in G1. Genes Dev 1993; 7:812–821.
55 Xiong Y, Zhang H, Beach D. D type cyclins associate with multiple protein kinases and the DNA replication and repair factor PCNA. Cell 1992; 71:505–514.
56 Ewen ME, Sluss HK, Sherr CJ et al. Functional interactions of the retinoblastoma protein with mammalian D-type cyclins. Cell 1993; 73:487–497.
57 Kato J, Matsushime H, Hiebert SW et al. Direct binding of cyclin D to the retinoblastoma gene product (pRb) and pRb phosphorylation by the cyclin D–dependent kinase CDK4. Genes Dev 1993; 7:331–342.
58 Bates S, Bonetta L, MacAllan D et al. CDK6 (PLSTIRE) and CDK4 (PSK–J3) are a distinct subset of the cyclin–dependent kinases that associate with cyclin D1. Oncogene 1994; 9:71–79.
59 Matsushime H, Quelle DE, Shurtleff SA et al. D–type cyclin–dependent kinase activity in mammalian cells. Mol Cell Biol 1994; 14:2066–2076.
60 Meyerson M, Harlow E. Identification of G1 kinase activity for cdk6, a novel cyclin D partner. Mol Cell Biol 1994; 14:2077–2086.
61 Tam SW, Theodoras AM, Shay JW et al. Differential expression and regulation of Cyclin D1 protein in normal and tumor human cells: association with Cdk4 is required for Cyclin D1 function in G1 progression. Oncogene 1994; 9:2663–2674.
62 Koff A, Cross F, Fisher A et al. Human cyclin E, a new cyclin that interacts with two members of the CDC2 gene family. Cell 1991; 66:1217–1228.
63 Koff A, Giordano A, Desai D et al. Formation and activation of a cyclin E-cdk2 complex during the G1 phase of the human cell cycle. Science 1992; 257:1689–1694.
64 Dulic V, Lees E, Reed SI. Association of human cyclin E with a periodic G1–S phase protein kinase. Science 1992; 257:1958–1961.
65 Tsai LH, Lees E, Faha B et al. The cdk2 kinase is required for the G1–to–S transition in mammalian cells. Oncogene 1993; 8:1593–1602.
66 Ohtsubo M, Theodoras AM, Schumacher J et al. Human cyclin E, a nuclear protein essential for the G1–to–S phase transition. Mol Cell Biol 1995; 15:2612–2624.
67 Resnitzky D, Reed SI. Different roles for cyclins D1 and E in regulation of the G1–to–S transition. Mol Cell Biol 1995; 15:3463–3469.
68 Rogers S, Wells R, Rechsteiner M. Amino acid sequences common to rapidly degraded proteins: the PEST hypothesis. Science 1986; 234:364–368.
69 Hadwiger JA, Wittenberg C, Richardson HE et al. A family of cyclin homologs that control the G1 phase in yeast. Proc Natl Acad Sci USA 1989; 86:6255–6259.
70 Tyers M, Tokiwa G, Nash R et al. The Cln3–Cdc28 kinase complex of S. cerevisiae is regulated by proteolysis and phosphorylation. EMBO J 1992; 11:1773–1784.
71 Clurman BE, Sheaff RJ, Thress K et al. Turnover of cyclin E by the ubiquitin–proteasome pathway is regulated by cdk2 binding and cyclin phosphorylation. Genes Dev 1996; 10:1979–1990.
72 Won KA, Reed SI. Activation of cyclin E/CDK2 is coupled to site–specific autophosphorylation and ubiquitin–dependent degradation of cyclin E. EMBO J 1996; 15:4182–4193.
73 Shuttleworth J, Godfrey R, Colman A. $p40^{MO15}$, a cdc2–related protein kinase involved in negative regulation of meiotic maturation of Xenopus oocytes. EMBO J 1990; 9:3233–3240.
74 Fesquet D, Labbe JC, Derancourt J et al. The MO15 gene encodes the catalytic subunit of a protein kinase that activates cdc2 and other cyclin–dependent kinases (CDKs) through phosphorylation of Thr161 and its homologues. EMBO J 1993; 12:3111–3121.
75 Solomon MJ, Harper JW, Shuttleworth J. CAK, the $p34^{cdc2}$ activating kinase, contains a protein identical or closely related to $p40^{MO15}$. EMBO J 1993; 12:3133–3142.
76 Poon RY, Yamashita K, Adamczewski JP et al. The cdc2–related protein $p40^{MO15}$ is the catalytic subunit of a protein kinase that can activate $p33^{cdk2}$ and $p34^{cdc2}$. EMBO J 1993; 12:3123–3132.

77 Darbon JM, Devault A, Taviaux S et al. Cloning, expression and subcellular localization of the human homolog of p40[MO15] catalytic subunit of cdk–activating kinase. Oncogene 1994; 9:3127–3138.
78 Levedakou EN, He M, Baptist EW et al. Two novel human serine/threonine kinases with homologies to the cell cycle regulating Xenopus MO15, and NIMA kinases: cloning and characterization of their expression pattern. Oncogene 1994; 9:1977–1988.
79 Tassan JP, Schultz SJ, Bartek J et al. Cell cycle analysis of the activity, subcellular localization, and subunit composition of human CAK (CDK–activating kinase). J Cell Biol 1994; 127:467–478.
80 Wu L, Yee A, Liu L et al. Molecular cloning of the human CAK1 gene encoding a cyclin–dependent kinase–activating kinase. Oncogene 1994; 9:2089–2096.
81 Fisher RP, Morgan DO. A novel cyclin associates with MO15/CDK7 to form the CDK–activating kinase. Cell 1994; 78:713–724.
82 Makela TP, Tassan JP, Nigg EA et al. A cyclin associated with the CDK–activating kinase MO15. Nature 1994; 371:254–257.
83 Feaver WJ, Svejstrup JQ, Henry NL et al. Relationship of CDK–activating kinase and RNA polymerase II CTD kinase TFIIH/TFIIK. Cell 1994; 79:1103–1109.
84 Serizawa H, Makela TP, Conaway JW et al. Association of Cdk–activating kinase subunits with transcription factor TFIIH. Nature 1995; 374:280–282.
85 Shiekhattar R, Mermelstein F, Fisher RP et al. Cdk–activating kinase complex is a component of human transcription factor TFIIH. Nature 1995; 374:283–287.
86 Leopold P, O'Farrell PH. An evolutionarily conserved cyclin homolog from Drosophila rescues yeast deficient in G1 cyclins. Cell 1991; 66:1207–1216.
87 Lahue EE, Smith AV, Orr–Weaver TL. A novel cyclin gene from Drosophila complements CLN function in yeast. Genes Dev 1991; 5:2166–2175.
88 Tassan JP, Jaquenoud M, Leopold P et al. Identification of human cyclin–dependent kinase 8, a putative protein kinase partner for cyclin C. Proc Natl Acad Sci USA 1995; 92:8871–8875.
89 Rickert P, Seghezzi W, Shanahan F et al. Cyclin C/CDK8 is a novel CTD kinase associated with RNA polymerase II. Oncogene 1996; 12:2631–2640.
90 Tamura K, Kanaoka Y, Jinno S et al. Cyclin G: a new mammalian cyclin with homology to fission yeast Cig1. Oncogene 1993; 8:2113–2118.
91 Wu L, Liu L, Yee A et al. Molecular cloning of the human CYCG1 gene encoding a G-type cyclin: overexpression in human osteosarcoma cells. Oncol Rep 1994; 1:705–711.
92 Horne MC, Goolsby GL, Donaldson KL et al. Cyclin G1 and cyclin G2 comprise a new family of cyclins with contrasting tissue–specific and cell cycle–regulated expression. J Biol Chem 1996; 271:6050–6061.
93 Okamoto K, Beach D. Cyclin G is a transcriptional target of the p53 tumor suppressor protein. EMBO J 1994; 13:4816–4822.
94 Zauberman A, Lupo A, Oren M. Identification of p53 target genes through immune selection of genomic DNA: the cyclin G gene contains two distinct p53 binding sites. Oncogene 1995; 10:2361–2366.
95 Skotzko M, Wu L, Anderson WF et al. Retroviral vector–mediated gene transfer of antisense cyclin G1 (CYCG1) inhibits proliferation of human osteogenic sarcoma cells. Cancer Res 1995; 55:5493–5498.
96 Bai C, Richman R, Elledge SJ. Human cyclin F. EMBO J 1994; 13:6087–6098.
97 Kraus B, Pohlschmidt M, Leung AL et al. A novel cyclin gene (CCNF) in the region of the polycystic kidney disease gene (PKD1). Genomics 1994; 24:27–33.
98 Nakamura T, Sanokawa R, Sasaki YF et al. Cyclin I: a new cyclin encoded by a gene isolated from human brain. Exp Cell Res 1995; 221:534–542.
99 Meyerson M, Enders GH, Wu CL et al. A family of human cdc2–related protein kinases. EMBO J 1992; 11:2909–2917.
100 van den Heuvel S, Harlow E. Distinct roles for cyclin–dependent kinases in cell cycle control. Science 1993; 262:2050–2054.
101 Hofmann F, Livingston DM. Differential effects of cdk2 and cdk3 on the control of pRb and E2F function during G1 exit. Genes Dev 1996; 10:851–861.

102 Tsai LH, Takahashi T, Caviness VS et al. Activity and expression pattern of cyclin–dependent kinase 5 in the embryonic mouse nervous system. Development 1993; 119:1029–1040.
103 Ohshima T, Ward JM, Huh CG et al. Targeted disruption of the cyclin–dependent kinase 5 gene results in abnormal corticogenesis, neuronal pathology and perinatal death. Proc Natl Acad Sci USA 1996; 93:11173–11178.
104 Zhang H, Xiong Y, Beach D. Proliferating cell nuclear antigen and p21 are components of multiple cell cycle kinase complexes. Mol Biol Cell 1993; 4:897–906.
105 Miyajima M, Nornes HO, Neuman T. Cyclin E is expressed in neurons and forms complexes with cdk5. Neuroreport 1995; 6:1130–1132.
106 Lew J, Beaudette K, Litwin CM et al. Purification and characterization of a novel proline–directed protein kinase from bovine brain. J Biol Chem 1992; 267:13383–13390.
107 Lew J, Huang QQ, Qi Z et al. A brain–specific activator of cyclin–dependent kinase 5. Nature 1994; 371:423–426.
108 Tsai LH, Delalle I, Caviness VS et al. p35 is a neural–specific regulatory subunit of cyclin–dependent kinase 5. Nature 1994; 371:419–423.
109 Ishiguro K, Kobayashi S, Omori A et al. Identification of the 23 kDa subunit of tau protein kinase II as a putative activator of cdk5 in bovine brain. FEBS Lett 1994; 342:203–208.
110 Uchida T, Ishiguro K, Ohnuma J et al. Precursor of cdk5 activator, the 23 kDa subunit of tau protein kinase II: its sequence and developmental change in brain. FEBS Lett 1994; 355:35–40.
111 Nikolic M, Dudek H, Kwon YT et al. The cdk5/p35 kinase is essential for neurite outgrowth during neuronal differentiation. Genes Dev 1996; 10:816–825.
112 Chae T, Kwon YT, Bronson R et al. Mice lacking p35, a neuronal specific activator of cdk5, display cortical lamination defects, seizures, and adult lethality. Neuron 1997; 18:29–42.
113 Pines J. The cell cycle kinases. Semin Cancer Biol 1994; 5:305–313.
114 Kobayashi H, Stewart E, Poon R et al. Identification of the domains in cyclin A required for binding to, and activation of, $p34^{cdc2}$ and $p32^{cdk2}$ protein kinase subunits. Mol Biol Cell 1992; 3:1279–1294.
115 Lees EM, Harlow E. Sequences within the conserved cyclin box of human cyclin A are sufficient for binding to and activation of cdc2 kinase. Mol Cell Biol 1993; 13:1194–1201.
116 Ducommun B, Brambilla P, Felix MA et al. cdc2 phosphorylation is required for its interaction with cyclin. EMBO J 1991; 10:3311–3319.
117 Endicott JA, Nurse P, Johnson LN. Mutational analysis supports a structural model for the cell cycle protein kinase p34. Protein Eng 1994; 7:243–253.
118 De Bondt HL, Rosenblatt J, Jancarik J et al. Crystal structure of cyclin–dependent kinase 2. Nature 1993; 363:595–602.
119 Jeffrey PD, Russo AA, Polyak K et al. Mechanism of CDK activation revealed by the structure of a cyclinA–CDK2 complex. Nature 1995; 376:313–320.

Chapter 4

Cyclin–CDK Regulation: Post–Translational Modification

Cyclins and cyclin–dependent kinases (CDKs) lie at the heart of the regulatory scheme that governs the progression of the cell cycle. The phase transitions are subject to the control of checkpoints, the biochemical sensing mechanisms that assay the environment of the cell. These checkpoints are in turn linked to cyclin–CDK complexes. The cyclin–CDK complex is therefore the ultimate arbiter of whether a cell is held in abeyance or whether it proceeds towards a proliferative destiny. Accordingly, the cyclin–CDK complex must be receptive to a variety of signals generated by intracellular and extracellular stimuli, synthesizing all available information to arrive at a proper decision. This is especially important if conditions are not appropriate for the cell to proliferate, for an improper decision to go forward can deregulate the cell cycle and potentially lead to a cancerous state or death.

By what means can cells use cyclins–CDKs to assimilate the wealth of information relayed by the cell's signaling pathways? As we have already seen, the most basic device by which the cell can pace successive cell cycles is inherent in the physical interaction of cyclins and CDKs. Because of the periodic rise and fall of cyclin levels versus the relative constancy of CDK levels, an interval of time must have passed after the destruction of cyclin in the previous cell cycle before enough new cyclin is synthesized to allow the formation of cyclin–CDK complexes; the cyclin level must stoichiometrically match the CDK level for the two proteins to stably interact. Since the binding of the cyclin partner is necessary for the CDK to be active, successive cell cycles must be separated by the length of time it takes the cell to synthesize sufficient cyclin. The consequence is that a mitosis cannot follow immediately after the previous mitosis. This guarantees the cell an interval of time in which it has the opportunity to grow, replicate DNA, and carry out other important cellular functions. Indeed, the promoters of many cyclin genes are phase–specific, ensuring synthesis of cyclins only at appropriate times.

The existence of an array of cyclin and CDK isoforms gives the cell the versatility to pace not only successive cell cycles but also successive phases of the cell cycle. The presence of certain cyclin–CDKs and the absence of others may very well serve as the primary indicators to the cell of what phase it resides in.[1,2] Active cyclin D– or cyclin E–CDK complexes correlate with G1 phase, whereas cyclin A– and cyclin B–CDK complexes correlate with G2 phase. It seems logical that each type of cyclin–CDK complex should curtail the activity of the previous cyclin–CDK and prepare the cell for the activation of the subsequent cyclin–CDK, thereby ensuring a proper sequence of cellular events.[3,4] For example, the recent observation that CDK2, which is responsible for the timing of DNA synthesis, is a required upstream positive regulator of cyclin B–CDK1 activity, which initiates mitosis, suggests an elegant mechanism by which mitosis is deferred until after DNA synthesis has taken place.[5]

Nevertheless, the physical interaction of cyclins and CDKs is too imprecise a device to regulate all aspects of the cell cycle. While the interaction ensures that events unfold at the proper pace and in the proper order, it is not sophisticated enough in and of itself to expeditiously respond to environmental stimuli. If a cell were to be exposed to a DNA–damaging agent, it would be necessary for the cell to arrest its passage through the cell cycle and make the appropriate repairs to its chromosomes. In one conceivable scenario, the cell would destroy its cyclin, precluding a situation in which active cyclin–CDK complexes could push the cell through a phase transition prematurely, and then resynthesize the cyclin after repairs were completed. But this drastic remove–and–replace scheme would be an inefficient process, as well as inconsistent with the observed pattern of continuous cyclin accumulation throughout the cell cycle.

A more viable hypothesis is that there exist regulatory factors in addition to cyclins and CDKs that influence cyclin–CDK complex kinase activity. One can imagine these factors playing various roles: physically blocking the interaction of cyclins and CDKs; allowing the cyclin–CDK complexes to form but physically inhibiting the kinase domains, whether directly or allosterically; phosphorylating or dephosphorylating key sites on the cyclins or CDKs in such a way as to prevent the formation of complexes or render complexes inactive; acting as transcriptional repressors to slow the production of cyclins and thereby the assembly of cyclin–CDK complexes. If exposed to a DNA–damaging agent, a cell could utilize one or more of these regulatory agents to inhibit the function of cyclin–CDK complexes, thereby suspending the cell cycle while preserving cyclin and CDK levels. When repair was complete, the cell would desist from using these negative regulators, restoring the ability of cyclins and CDKs to form active complexes. Motion through the cell cycle would immediately resume.

Not surprisingly, it has been demonstrated that mammalian cells do indeed employ such factors for both positive and negative regulation of cyclin–CDK complexes. At the present time we are sure of two mechanisms by which the regulation is enacted: physical inhibition of the complexes, and post–translational modification of the CDK subunit. The characterization of CDK inhibitors (CKIs) is a recent occurrence, this family of proteins having been discovered only in the past few years. The evidence that CDKs are subjected to modification by other kinases, however, has been accruing since the identification of the *CDC28* and *cdc2*$^{+}$ cell cycle control genes in *Saccharomyces cerevisiae* and *Schizosaccharomyces pombe* in the 1970s. The original screens for cell division cycle mutants in these yeast systems uncovered genes that subsequently have been found to encode CDK regulators. We shall therefore briefly review the relevant developments in yeast genetics before turning our attention to the specifics of post–translational CDK modification. We will take up the subject of CKIs in the next chapter.

CDK1 is Subject to Modification During the Yeast Cell Cycle

In the context of the cell cycle, the onset of mitosis is a remarkably rapid process in which a host of events unfold contemporaneously. The overarching questions for the next few sections are these: how does the cell know exactly when to make the abrupt transition from G2 phase to M phase, and what kind of mechanism can provoke the large number of simultaneous events that occur immediately after the transition is made?

In the original studies which characterized the cyclins, it was noted that the onset of mitosis corresponded with the peaking of cyclin levels.[6] This observation suggested to investigators that the cell's accumulation of cyclin was required to surpass a specific threshold level at which mitosis would be induced. Upon the identification of maturation–promoting factor (MPF) as a cyclin B–CDK1 complex, this hypothesis matured into a model in which cyclin B stoichiometrically activated the complex only after surpassing the threshold level.[7]

However, it must be remembered that the initial cyclin studies were performed with embryonic cells, and that the embryonic cell cycle omits the gap phases in the interest of generating a large number of nuclei in a short period of time. How the threshold model would apply to normally dividing somatic cells, in contrast to cells with an accelerated cell cycle, should be taken into consideration. In the full–length cell cycle of somatic cells, the extended pace allows cyclin

B to accumulate over a longer period of time. The putative threshold level should be reached a set interval after the start of G2 phase, and at that point cyclin–CDK complexes should appear. If we assume that the presence of cyclin–CDK complexes will induce M phase, then we arrive at the conclusion that G2 phase should be of a more or less fixed length. But we should expect the length of G2 phase to be variable, dependent on the appropriateness of conditions for cell division. How, then, can the dramatically rapid cellular changes associated with the onset of mitosis be precisely timed, i.e., either delayed or expedited in response to environmental signals, if MPF activity is turned on by the cyclin level crossing a defined threshold?

The response to this quandary was to postulate that the binding of cyclin B to CDK1 was not sufficient to activate the complex but that another factor was necessary. With the support of experimental evidence, a new model was formulated in which "pre–MPF," inactive cyclin B-CDK1 complexes, was amassed well before the end of G2 phase.[8-13] At the G2/M phase transition, a sudden modification of the pre–MPF rendered it into active MPF, allowing the rapid induction of mitosis (Fig. 4.1). In a sense there remained a threshold to be passed, for enough pre–MPF had to be synthesized in order to generate sufficient MPF to provoke entry into M phase. But the crucial event was a post–translational modification enacted through some hitherto unknown mechanism. Interestingly, the identity of this mechanism was provided not by biochemical means but by a genetic line of evidence established in yeast systems.

During the years when *cdc2*$^{+}$ and *CDC28* were identified and characterized in *S. pombe* and *S. cerevisiae*, a regulatory network acting on these yeast homologues of CDK1 was pieced together as well.[14] The initial search for cell division cycle mutants in *S. pombe* identified the *cdc25*$^{+}$ gene that, when mutated, caused cells to arrest in G2 phase.[15] Moreover, when *cdc25*$^{+}$ was overexpressed the onset of mitosis was accelerated in a dose–dependent manner, implicating the gene product as a mitotic inducer.[16-18] A *cdc25*$^{+}$ homologue in *S. cerevisiae*, *MIH1*, was identified by functional complementation;[19] *cdc25*$^{+}$ homologues were also found in *Drosophila* fruit flies and humans.[20-22]

A search for mitotic regulatory genes uncovered mutant *S. pombe* cells that were abnormally small in size throughout the cell cycle, a phenotype that was consistent with the additional observation that G1 phase was prolonged and G2 phase was shortened. Evidently the G2 regulatory machinery was short–circuited, allowing the cells to undergo mitosis before they were of proper size; the miniature daughter cells required extra time to grow to sufficient size to pass the G1/S threshold. This "wee" phenotype was found to be a consequence of mutations in the aptly named *wee1*$^{+}$ gene.[23-25] Overexpressed *wee1*$^{+}$ caused cells to assume a large, elongated phenotype, reminiscent of cell division cycle mutants, in a dose–dependent manner.[26] *wee1*$^{+}$ was clearly a mitotic inhibitor, the functional oppo-

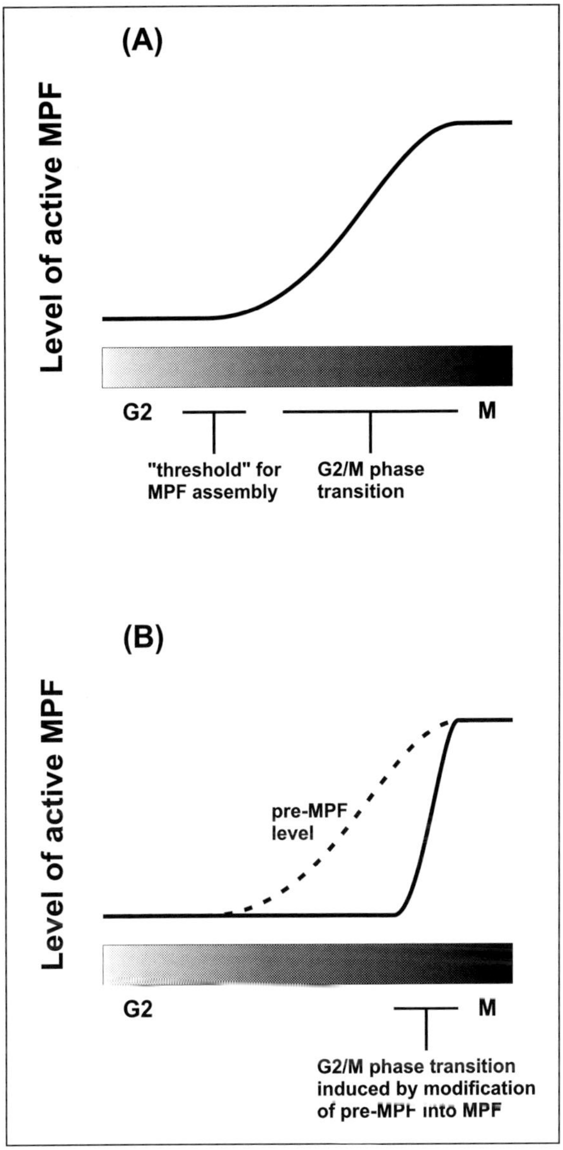

Fig. 4.1. Diagrams depicting two alternative models of MPF activation and the cell's passage through the G2/M phase transition. (A) The cell's accumulation of cyclin B surpasses a stoichiometric "threshold" level during G2 phase, allowing the assembly of active cyclin B–CDK1 (MPF) complexes. The continuing synthesis of cyclin B allows for a gradual increase in the level of active MPF until the start of M phase, when the cyclin B level reaches its peak. This model has two related weaknesses. First, there is a lengthy interval at the end of G2 phase, before the start of M phase, in which active MPF is present in the cell and free to phosphorylate downstream targets. We should expect to see various processes preparatory for mitosis staggered throughout this interval. This expected pattern contradicts what is actually observed in the cell; the onset of mitosis is a dramatic process and many events unfold more or less simultaneously. Second, since this model says only that events associated with the onset of mitosis will commence upon the cell's passing the "threshold," there are no provisions for the cell to halt or speed its advance into mitosis in response to environmental signals. (B) In the second model, there remains a "threshold" for the formation of cyclin B–CDK1 complexes. But these complexes do not constitute active MPF; rather, they are an inactive form of MPF that is designated "pre–MPF." A specific modification event must be executed by the cell in order to transform pre–MPF into MPF *en masse*. Because the conversion of pre–MPF to MPF takes place very rapidly, the G2/M phase transition occurs in a relatively short, clearly–defined interval. The cell can adjust the timing of the G2/M phase transition in response to environmental signals by adjusting the timing of the modification event.

site of *cdc25⁺*. A gene with behavior similar to *wee1⁺* in human cells was identified as a possible *wee1⁺* homologue.[27]

Interestingly, the wee phenotype was shown in some cases to be linked to mutations in *cdc2⁺* (designated as *cdc2–wee* mutants) rather than *wee1⁺*.[24,25] The implication was that *wee1⁺* exerted its inhibitory effect directly through *cdc2⁺*; in such a model, *wee1⁺* mutants failed to inhibit *cdc2⁺*, and *cdc2–wee* mutants failed to be inhibited by *wee1⁺*. Upon characterization of the *cdc2–wee* mutants, they were found to comprise two different classes of *cdc2* alleles.[24,28] In the first class, not surprisingly, the *cdc2* alleles gave rise to the wee phenotype irrespective of the presence or absence of *wee1⁺*. In contrast, the second class of *cdc2* alleles gave the phenotype independently of the presence of *cdc25⁺*. These findings suggested that the gene products of *wee1⁺* and *cdc25⁺* interacted with $p34^{cdc2}$ via two distinct but convergent pathways, antagonistic to one another.

That *cdc25⁺* and *wee1⁺* functioned in parallel in the regulation of *cdc2⁺* was demonstrated by combinatorial studies of the two genes (Fig. 4.2). Conditional mutants in *cdc25⁺*, arrested in G2 phase and with an elongated phenotype, were rescued by mutations in *wee1⁺*, demonstrating that they had opposing functions.[16,17] When *wee1⁺* was deleted and *cdc25⁺* was overexpressed in the same system, cell division was so accelerated that the cells were too small to survive mitosis—the so–called "mitotic catastrophe."[17] The synergistic effect of the manipulation of both genes implied that they constituted a positive/negative pair enjoined in a single regulatory device.

An elegant explanation of the effects of *wee1⁺* and *cdc25⁺* was provided when $p34^{cdc2}$ was observed to vary in its phosphorylation state during the *S. pombe* cell cycle, specifically at the sites Tyr15 and Thr167, two amino acids conserved among all known $p34^{cdc2}$ homologues.[29-31] At the start of M phase, Tyr15 was dephosphorylated; in contrast, Thr167 was phosphorylated. Importantly, removal of the phosphate group from Tyr15 was seen to be concurrent with, as well as required for, the onset of mitosis.[29,32] Moreover, phosphorylation of Thr167 was found to be involved with the binding of cyclin B to $p34^{cdc2}$; indeed, it was necessary for $p34^{cdc2}$ to have kinase activity.[30,33] Evidently the kinase was subject to regulation by one or more protein kinase/phosphatase pairs.

Biochemical characterization of the gene product of *cdc25⁺*, which hereafter will be referred to as CDC25, revealed it to be the protein phosphatase responsible for the dephosphorylation of Tyr15.[34-40] Confirmation came in the finding that *cdc25⁺* expression increased during interphase and peaked during mitosis in a periodic fashion.[41-43] The *wee1⁺* gene product, hereafter called Wee1, was found to harbor both serine/threonine and tyrosine kinase activities, implicating it as the protein kinase that phosphorylated Tyr15.[44-46] Although the theoretical possibility existed that Wee1 phosphorylated Thr167 as well, this was later shown not to be the case.

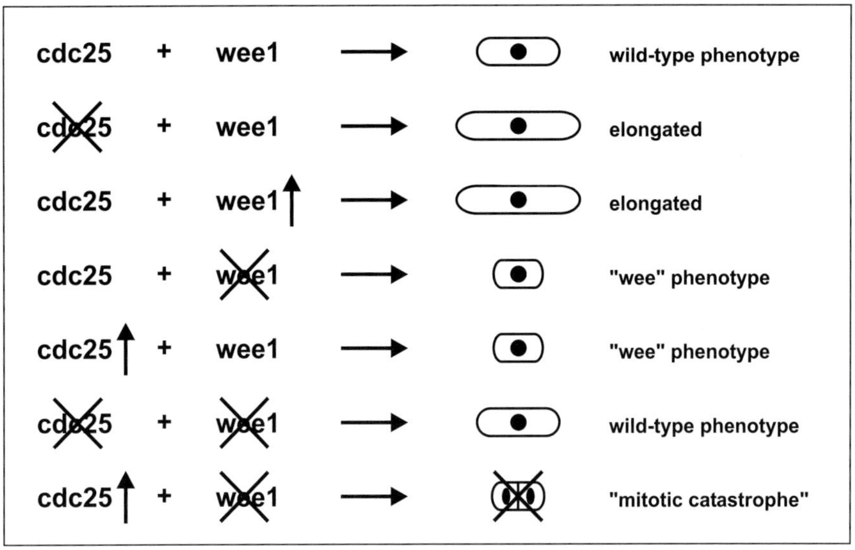

Fig. 4.2. Diagram depicting the results of experiments with *Schizosaccharomyces pombe* cells in which the *cdc25*$^+$ gene product and/or the *wee1*$^+$ gene product is altered. A loss–of–function cdc25 mutant induces a G2 phase arrest, allowing the cell to grow for an extended period of time and thereby attain an elongated phenotype. The same result is achieved by the overexpression of wee1 in the cell. In contrast, a loss–of–function wee1 mutant causes the assumption of a "wee" phenotype in which the cell is abnormally small in size due to an accelerated G2 phase and mitosis; overexpression of wild–type cdc25 yields a similar result. Mutations in both cdc25 and wee1 allow the cell to retain its wild–type phenotype, whereas mutation of wee1 alone coupled with overexpression of cdc25 causes the so–called "mitotic catastrophe," in which G2 phase is shortened to such an extent that the subsequent mitosis yields progeny cells that are too small to survive. The interpretation of these experiments: cdc25 and wee1 interact with $p34^{cdc2}$ via two convergent, functionally opposite pathways; because of the synergistic effect of altering both proteins in the last experiment, it can be further concluded that they constitute a positive/negative pair antagonistically regulating $p34^{cdc2}$.

The point must be made that CDC25 as a regulator of cyclin B–CDK1 is vulnerable to the same criticism as cyclin B as a regulator of CDK1—because of its periodic behavior during the cell cycle, it fails to provide an answer in and of itself to the question of how mitosis is precisely timed in accordance with environmental signals. By a similar line of reasoning as the arguments given at the beginning of this chapter, it seems that CDC25 should itself be regulated in the same manner as CDK1. Indeed, evidence was soon found that CDC25 is subject to regulation by yet another protein kinase/phosphatase pair.[47] Complicating the situation even further was the suspicion that Wee1 was also under the control of a protein kinase/phosphatase pair.[48,49]

From these studies emerged a picture in which the interplay of protein kinases and phosphatases was central to CDK1 (and therefore mitotic) regulation. Accordingly, recent years have witnessed an enormous interest in identifying kinases and phosphatases that double as cell cycle regulators.

Protein Kinases and Phosphatases

The current model of CDK1 regulation features interwoven feedback loops mediated by a number of protein kinases and phosphatases. While many of the relevant proteins are yet to be firmly identified, enough have been characterized to the extent that we can piece together an informative picture.

CDK1 is inactive when it is phosphorylated at the Tyr15 residue in *S. pombe*, and when it is phosphorylated at both the Thr14 and Tyr15 residues in vertebrate cells. These residues, near the N–terminus of the molecule, are located in the ATP–binding domain; when they are phosphorylated they seem to interfere with the CDK's kinase activity not by impeding the binding of nucleotide but rather by inhibiting the subsequent catalysis.[50-52] Interestingly, the phosphorylation of these residues increases gradually through interphase. When coupled with the observation that these residues are subject to phosphorylation only when CDK1 is bound to cyclin, it appears that the phosphorylation of CDK1 is correlated with its binding to cyclin.[12,31,50,53] This is consistent with the theory that a large pool of "pre–MPF" is synthesized prior to the onset of mitosis. It is easy to envision that the binding of cyclin to CDK1 causes a conformational change in CDK1 that exposes Thr14 and Tyr15, at which time these residues are immediately phosphorylated. This mechanism ensures that when MPF is initially assembled its activity is instantly suppressed, preventing a catastrophic premature mitosis.

The available evidence suggests that, in yeast, Wee1 and its close cousin Mik1 are primarily responsible for the negative regulation of CDK1 through the phosphorylation of Tyr15.[44,46,54-56] In vertebrates, only one Wee1–related protein kinase that phosphorylates Tyr15 has been found.[27,57] Because Wee1 is a dual–specificity kinase that can phosphorylate both tyrosine and threonine residues,[46] one might suspect that it also phosphorylates the adjacent Thr14. Interestingly, vertebrate Wee1 is capable of phosphorylating only the Tyr15 residue, implying the existence of an independent protein kinase responsible for Thr14 phosphorylation.[58,59] (The *S. pombe* version of Wee1 apparently can phosphorylate the Thr14 residue of yeast CDK1, but the significance is unclear since Thr14 does not appear to play a major role in CDK1 regulation in yeast, unlike vertebrates.[60]) Indeed, a kinase capable of phosphorylating Thr14 has been found in *Xenopus laevis* (identified as Myt1), and comparable kinase activity has been detected in human cells; further, the kinase appears to be of dual specificity, acting on both the Thr14 and Tyr15 residues.[61-63] It has been shown that phosphorylating only one of the two residues still significantly impairs CDK1 activity.[59,64-66] As to why there exist two distinct kinases with almost redundant functions, one may speculate that they constitute separate pathways through which different upstream regulators exert their influence on CDK1; the added complexity would allow for a more diverse range of inputs to be "processed" by CDK1. Another possibility is that other CDKs, which presumably are subject to the same regulatory phosphorylations as CDK1, are differentially acted upon by the pair of kinases.

Wee1 is itself regulated by protein kinases and phosphatases. In *S. pombe*, the protein kinase Nim1/Cdr1 inhibits Wee1 activity by phosphorylating a residue in the catalytic domain near the C–terminus.[48,49,67-70] A number of other kinases, including Wis1 and Win1, have also been implicated in acting upon Wee1.[71-73] No kinases that phosphorylate Wee1 have yet been identified in vertebrates, but in several species Wee1 has been observed to oscillate in its phosphorylation state in a cell–cycle–dependent manner, becoming hyperphosphorylated (and thereby inactive) at the start of M phase.[74-78] This observation confirms that Wee1 is of particular importance in inhibiting CDK1, that removing Wee1 activity via phosphorylation allows CDK1 to drive the cell into M phase. (Also implicit in this observation is that Wee1 is engaged in a "tug–of–war" with an antagonistic phosphatase, for upon Wee1's inactivation the Thr14 and Tyr15 residues of CDK1 must somehow be dephosphorylated before MPF can become active and initiate mitosis. This will be addressed in more detail below.) Notably, vertebrate Wee1 seems to be inhibited by phosphorylation of residues near the N-terminus, demonstrating that Wee1 is subject to negative regulation by more than one pathway.[74,75] Also interesting is the finding that MPF directly or through an intermediary kinase phosphorylates Wee1, possibly establishing a positive feedback loop that generates more active MPF.[75,77]

Wee1 is restored to its hypophosphorylated state after mitosis is complete and cyclins A and B have been destroyed, allowing the kinase to once again inhibit any newly synthesized MPF complex. The phosphatase(s) responsible for counteracting yeast Nim1 and its vertebrate counterparts are as of yet unidentified. Curiously, two *S. pombe* phosphatases, Pyp1 and Pyp2, have been shown to inhibit mitosis by strengthening Wee1's inhibition of CDK1.[79,80] But they evidently do not act directly upon Wee1, working instead through some unknown indirect means.[81]

CDC25 is the dual–specificity phosphatase primarily responsible for dephosphorylating both the Thr14 and Tyr15 residues of CDK1, making it the antagonist to Wee1.[35-40,82,83] The catalytic domain of CDC25 is located at the C–terminus.[19,37,84] Mammalian cells have three versions of CDC25, called CDC25A, CDC25B and CDC25C, which all share the crucial catalytic domain.[85] The regulatory domain of CDC25 is located at the N-terminus; CDC25A, CDC25B and CDC25C diverge in the constitution of this domain.[85] Accordingly, the three isoforms are constrained to act at different times during the cell cycle. CDC25A activity peaks at the G1/S phase transition, and CDC25C peaks at the G2/M phase transition; the behavior of CDC25B is presently unclear.[22,84-87]

In yeast, CDC25 is expressed in a cyclical fashion during the cell cycle, peaking at the onset of mitosis.[41-43] In contrast, the vertebrate isoforms of CDC25 are maintained at more or less constant levels during the cell cycle, and their activity is post–translationally modulated by the action of kinases and phosphatases.[47,86,88,89] CDC25C is activated at the G2/M phase transition by phosphorylation at several residues in the N–terminal regulatory domain; moreover, the kinase responsible for the phosphorylation appears to be none other than cyclin B–CDK1.[47,90-94] Similarly, CDC25A appears to be phosphorylated at the G1/S transition by cyclin E-CDK2.[95] The implication of these findings is that CDC25 is part of a positive feedback loop by which CDK1 (as well as CDK2 and possibly other CDKs) stimulates its own activation. There is also evidence that, besides being phosphorylated by cyclin–CDK complexes, CDC25 is acted upon by at least one other kinase. It has been hypothesized that the alternate kinase or kinases jumpstart CDK1 activation by initially phosphorylating CDC25, after which CDK1 activation sustains itself by carrying out the larger part of the CDC25 phosphorylation.[92,96,97] Recent studies suggest that the Raf1 and Myc proto–oncogene products may directly activate CDC25 by phosphorylation and transcriptional transactivation respectively, behavior that is consistent with their oncogenic potential.[98,99]

As with Wee1, CDC25 may have partners with overlapping redundant function. Two genes have recently been isolated in *S. pombe* by their ability to suppress a mutation in the *cdc25*$^+$ gene; they encode two phosphatases, Pyp3 and Stp1, that are capable of dephosphorylating CDK1.[100-103] Highly conserved mammalian homologues of Stp1 have been identified, but their function is yet to be clarified.[101]

There is a growing body of evidence that indicates that members of two classes of mammalian serine–threonine phosphatases, PP1 and PP2A, play an inhibitory role in mitotic control, although no individual phosphatase has been definitively assigned a position in the pathways impinging on CDK1.[47,90,93,104-117] It has been proposed that PP1 and/or PP2A phosphatases act upstream of CDC25, inactivating it through intermediaries or even directly dephosphorylating it, countering the influence of CDK1 and other CDC25–activating kinases. Other phosphatases may act as antagonists to Wee1–inactivating kinases. It is probable that the array of phosphatases utilized in CDK1 regulation varies widely among different species.

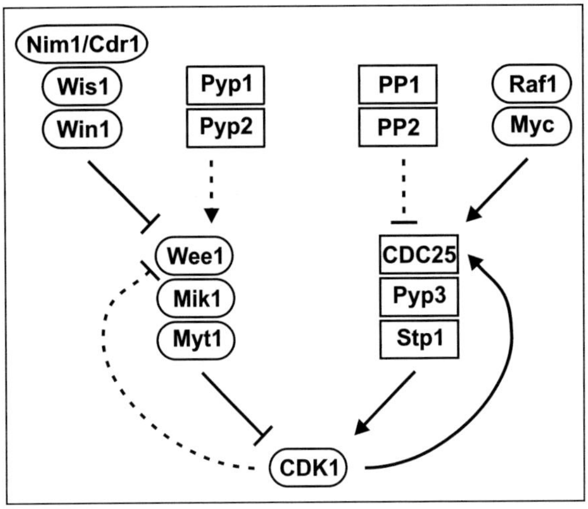

Fig. 4.3. Diagram depicting the regulatory network of protein kinases and phosphatases upstream of CDK1. Names in bubbles designate kinases, and names in boxes designate phosphatases; arrows indicate an activation of the target protein, and bars indicate an inhibition of the target protein; solid lines indicate a direct interaction between the two proteins, and dotted lines indicate that it is presently unclear whether the interaction is direct or through intermediary proteins. Please note that this figure was compiled using data from a number of species and does not reflect the regulatory network in the cells of any one system.

Figure 4.3 portrays the kinase/phosphatase network upstream of Wee1 and CDC25 as it is presently understood.

Regulatory Phosphorylation as a Model for Mitotic Control

We now have a viable response to the question of how the activity of one molecular complex, MPF, give rises to the abrupt profusion of cellular events that marks the transition from G2 phase to M phase. We also have arrived at an elegant explanation for the classic experiments that first identified MPF: when a small amount of MPF extract from "mature" oocytes was injected into "immature" oocytes, the latter became mature and generated large amounts of MPF; a small amount of extract from these newly–matured ooctyes could activate another set of immature oocytes, and the process could be repeated again and again (see chapter 2).[9,10,118-120] When even the smallest amount of active MPF is introduced into a late G2–phase cell, either by direct injection or by CDC25 dephosphorylation of endogenous pre–MPF, it dramatically self–amplifies via positive

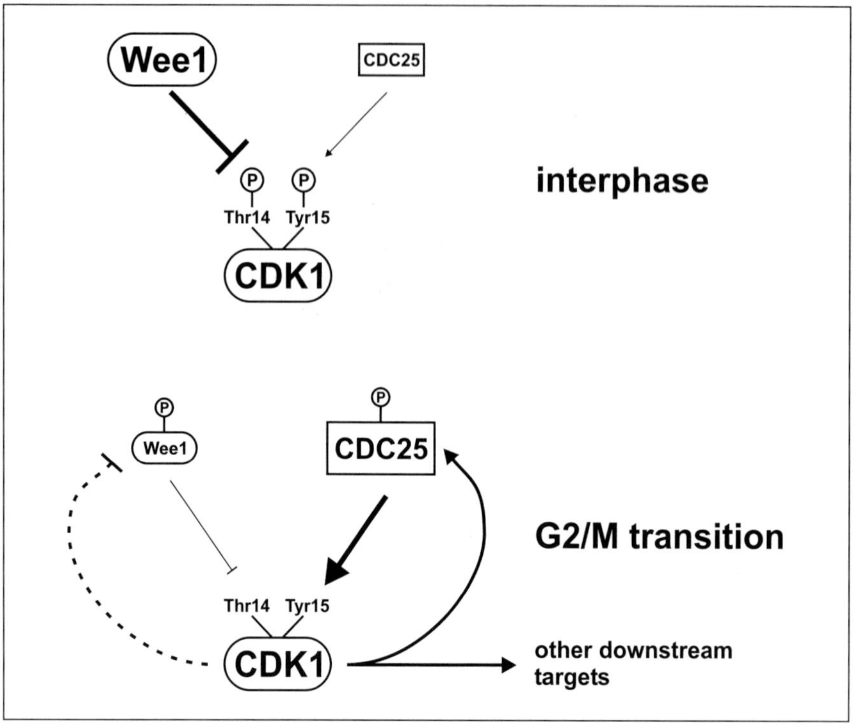

Fig. 4.4. Diagrams depicting the interactions of Wee1 (and its relatives), CDC25, and CDK1 during the cell cycle. In interphase, both Wee1 and CDC25 are predominantly unphosphorylated; this has the effect of rendering Wee1 mostly active and CDC25 mostly inactive. The Wee1 kinase has primacy in the "tug–of–war" with the CDC25 phosphatase and thereby maintains the phosphorylation of CDK1 on the Thr14 and Tyr15 residues, nullifying CDK1 activity. At the G2/M transition the "tug–of–war" sways in favor of CDC25. CDC25 is activated via phosphorylation by upstream regulators; Wee1 is similarly phosphorylated with the opposite effect of losing its kinase activity. CDK1 is dephosphorylated by CDC25 and thereby becomes active, capable of phosphorylating downstream targets and instigating the onset of mitosis. CDK1 also directly phosphorylates CDC25 and brings about the phosphorylation of Wee1 (whether directly or indirectly is not known), thus initiating feedback loops that reinforce the preponderance of CDC25 activity over that of Wee1 and the consequent dephosphorylation/activation of CDK1.

feedback loops. Within a very short time a large pool of active MPF is generated from pre–MPF, setting into motion the seemingly explosive chain of events associated with the onset of mitosis.

Taking a global view of CDK1 regulation, we are poised to understand how environmental signals exert their influence on mitotic control through pathways ultimately converging upon CDK1 (Fig. 4.4). A delicate balance is drawn between Wee1 and its partners on one side and CDC25 and its partners on the other, the former braking the cell cycle by inhibiting CDK1, the latter accelerating it through stimulation of CDK1.[121] Throughout G2 phase Wee1 holds the upper hand, keeping the Thr14 and Tyr15 residues of CDK1 in an unceasing phosphorylated state. The protein kinases and phosphatases upstream of Wee1 and CDC25 are nevertheless relaying environmental signals into the regulatory machinery by modulating the two regulators. Just before the G2/M phase transition, Wee1 activity is reduced by the action of Nim1 and other mechanisms; the phosphatases acting on CDC25 relax their suppression of its activity. Thus the balance is swayed slightly in favor of CDC25, and a few pre–MPF complexes are dephosphorylated. The MPF self–amplifies by engaging in coordinated feedback loops: stimulating CDC25 by phosphorylation (and perhaps by inhibiting the phosphatases upstream of CDC25) and inactivating Wee1 either directly or through intermediary kinases. CDC25 quickly overwhelms Wee1's antagonism, decisively dephosphorylating the Thr14 and Tyr15 residues of CDK1, which drives the cell into M phase. After mitosis ends and MPF is degraded, the CDC25 feedback loop ceases and the balance swings back in favor of Wee1. As cyclin is synthesized anew through the early part of interphase, Wee1 reestablishes its restraint on CDK1 through phosphorylation.

CDK1 regulation is a prototypical case of unstable equilibrium. Either Wee1 or CDC25 predominates; either the totality of CDK1 is phosphorylated at Thr14 and Tyr15 or it is not phosphorylated; either MPF is fully active or it is inactive. The feedback loops ensure that there can be no intermediate states. It is to the cell's advantage to exist at one extreme or the other, for an intermediate state means that a decision made by the cell concerning division is only partially supported by the proliferative machinery. Any ambivalence can lead to a mitotic catastrophe or, worse, a loss of dependence of one phase of the cell cycle on the proper completion of that preceding it—cell cycle deregulation in which there lies the danger of cancer.

Having described this elegant model of CDK1 regulation, it should be noted that the model does not universally represent mitotic control in all species and systems. Curiously, the phosphorylation and dephosphorylation of Tyr19 of *S. cerevisiae* CDK1, the residue equivalent to Tyr15 of *S. pombe* CDK1, does not have a crucial role in the regulation of the G2/M phase transition.[122,123] In goldfish oocytes, Tyr15 is not phosphorylated (and consequently never dephosphorylated) after CDK1 binds to cyclin B.[124] Further, Thr14/Tyr15 regulation is absent during the abbreviated cell cycles of early *Drosophila* embryonic cells; only after the 14th cycle, when full–length cell cycles with G2 phase first appear, does inhibitory phosphorylation of CDK1 begin.[125] These exceptions imply that there are additional mechanisms involved in mitotic regulation that supersede the Wee1/CDC25 pathways; it is conceivable that the Wee1/CDC25 network is used primarily as an "emergency" response and may not be relevant during "normal" asynchronous proliferation. To date at least one potential alternative mechanism has been well described. Interestingly, this mechanism implicates a novel cyclin–CDK complex as a significant participant in cyclin B-CDK1 regulation.

A Cyclin–CDK Complex Regulates MPF

Along with Thr14 and Tyr15, another threonine residue of CDK1 (Thr161 in *X. laevis*, Thr167 in *S. pombe*; for simplicity's sake we will hereafter refer to this residue as Thr161) is differentially phosphorylated during the extent of the cell cycle.[30,31,126] In contrast to the first two residues, Thr161 is found in an unphosphorylated state in early interphase and in a phosphorylated state at the G2/M phase transition. Whereas Thr14 and Tyr15 phosphorylation are inhibitory in mitotic control, Thr161 phosphorylation is stimulatory; indeed, the latter is requisite for CDK1 activity.[30,50,126] As with Thr14 and Tyr15, the binding of cyclin B to CDK1 seems to be crucial for the phosphorylation of Thr161; conversely, Thr161 phosphorylation appears to strengthen cyclin binding to CDK1.[30,31,33,126-128] Unlike Thr14 and Tyr15, which are located at the N–terminus and when phosphorylated are thought to interfere with the transfer of phosphate from ATP to the substrate, the Thr161 residue is buried in the T–loop, the structure that blocks access to the active site and prevents the substrate from coming into proximity to bound ATP (see chapter 3).[51,52,129] The binding of the cyclin to the CDK causes a displacement of the T–loop, which has several significant consequences: the entrance to the active site is opened, allowing ATP and the substrate to come into contact; new interactions between residues on the cyclin and the CDK are formed; and Thr161 becomes exposed to the environment and thus amenable to phosphorylation.[52] The phosphorylation of Thr161 makes the active site even more accessible by locking in the displacement of the T–loop and stabilizing the cyclin–CDK complex.[130]

The agent responsible for Thr161 phosphorylation is the aptly named CDK–activating kinase (CAK). CAK's catalytic component has been identified as the CDK–related MO15 protein kinase.[131-139] Subsequent analysis has revealed that another subunit of CAK is a member of the cyclin family of proteins, cyclin H.[140,141] In most species in which CAK has been characterized, the factor is composed of homologues of both MO15 and cyclin H.[137,140-145] CAK thus appears to conform to the cyclin–CDK archetype, and accordingly MO15 is now referred to as CDK7 (see chapter 3). CAK represents the first demonstration of a cyclin–CDK acting upon another cyclin–CDK. Indeed, the finding that CDK7 is itself subject to phosphorylation on a threonine residue (Thr170 in human CDK7, Thr176 in *X. laevis*; this is the residue homologous to Thr161 of CDK1) and that this phosphorylation stimulates its kinase activity suggests the possibility of the cyclin–CDK cascade as a novel paradigm of signal transduction.[140,142,146]

In a survey of vertebrate species, the CAK level and expression are observed to be constant throughout the cell cycle.[135,138,146,147] Consequently it is uncertain how important a role Thr161 phosphorylation plays as a regulatory device through which environmental signals can influence CDK1 activity. While Thr161 dephosphorylation of necessity must occur after mitosis in order to "reset" CDK1, it is not clear whether such a dephosphorylation might have an inhibitory role before mitosis. No phosphatase antagonist to CAK has yet been identified, although recent studies have uncovered two possible candidates for the role, a PP2A phosphatase called INH[109,117] and a novel CDK–binding phosphatase called KAP.[148-150] It is conceivable that phosphatases prevent CDK1 activation by either reversing Thr161 phosphorylation or by inhibiting CDK7 by dephosphorylation, but there is not yet compelling evidence that this actually occurs.

Many questions concerning CAK are unresolved, and recent findings have only complicated the issue. It appears that activation of CAK can be achieved by either of two seemingly unrelated mechanisms: the phosphorylation of the CDK7 subunit of CAK on Thr160/167 by an as–of–yet unidentified CAK–activating kinase (CAKAK),[140,142,146] or the association of CDK7 and cyclin H in a complex along with an assembly factor—a novel protein named MAT1 ('menage a trois')—without the need for CDK7 phosphorylation.[151-154] CDK7 and cyclin H have been identified as essential components of the transcription factor TFIIH, implicating CAK in dual roles in cell cycle regulation and transcriptional regulation.[155-157] Yet it is far from clear whether cyclin H-CDK7 is equivalent to CAK in all species. Kin28, an *S. cerevisiae* kinase that is highly homologous to CDK7, apparently is a component of yeast TFIIH but does not possess CAK activity.[158] Conversely, a CAK enzyme recently cloned from *S. cerevisiae* bears homology to CDKs but possesses full kinase activity in monomeric form, i.e., without a cyclin partner, and is not a component of TFIIH.[159,160] The significance of the alternative functions, identities, and activation mechanisms of CAK and/or cyclin H-CDK7 in the context of cell cycle regulation remains to be elucidated.

A General Schematic for CDK Regulation by Post–Translational Modification

The model of CDK1 regulation we have now examined in depth is at least superficially applicable to the entire CDK family. While specific pathways, phosphorylation states, etc., may be unique to individual CDKs, the core regulatory framework seems to be conserved among most CDKs. For instance, it has been demonstrated that CDK2 activity is negatively regulated via phosphorylation of

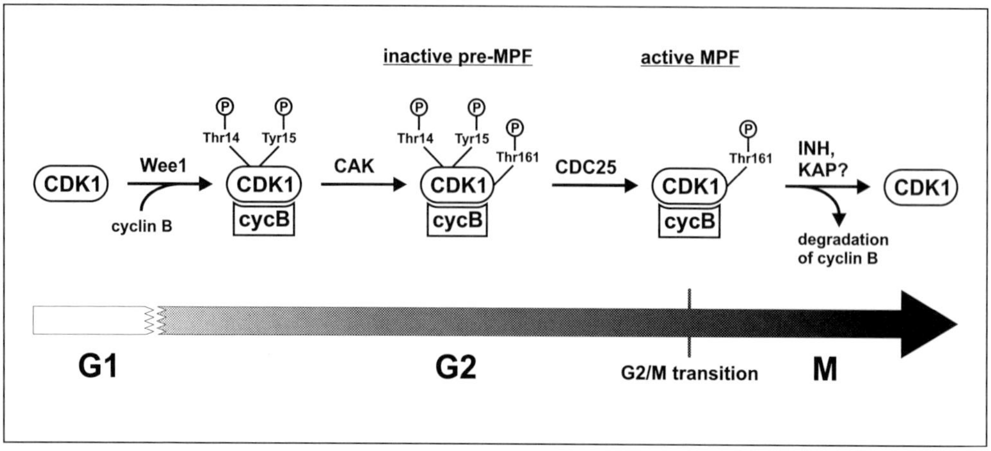

Fig. 4.5. Schematic illustration of the model of activation and inactivation of MPF complexes during the cell cycle. In G1 phase, only CDK1 is present in appreciable quantities, and it exists in a completely unphosphorylated state. By G2 phase, enough cyclin B has been synthesized to surpass the threshold for pre–MPF assembly. Concurrent with the binding of cyclin B to CDK1 is the phosphorylation of the Thr14 and Tyr15 residues of CDK1 by Wee1 and its relatives, assuring that the complex remains inactive. Shortly thereafter (or perhaps simultaneously) CDK1 is phosphorylated on the Thr161 residue by CAK, which in most species appears to be a cyclin H-CDK7 complex. When conditions are appropriate for mitosis, CDC25 will dephosphorylate the Thr14 and Tyr15 residues and render the cyclin B–CDK1 complex into active MPF. MPF drives the cell through the G2/M phase transition and initiates mitotic processes. During M phase, cyclin B is degraded through a ubiquitin–mediated pathway; through a process that is unclear, the Thr161 residue of CDK1 is dephosphorylated (possibly by the phosphatases INH and KAP). MPF activity is thereby abrogated. The cell cycle begins anew with unphosphorylated CDK1.

the Thr14 and Tyr15 residues, and both CDK2 and CDK4 are positively regulated via phosphorylation of Thr160 and Thr172 respectively (the residues homologous to Thr161 of CDK1).[132-134,137,161-164] All available evidence suggests that every CDK conforms to these regulatory principles.

We therefore present a generalized schematic for CDK post–translational modification, as portrayed in Figure 4.5. There are three devices through which a CDK's kinase activity can be positively regulated: (1) binding of a cyclin partner, (2) dephosphorylation of two residues at the N-terminus of the protein, corresponding to Thr14 and Tyr15 in CDK1, and (3) phosphorylation of a threonine residue in the T-loop, corresponding to Thr161 in CDK1. Inversely, negative regulation can be achieved through (1) degradation of the cyclin partner, (2) phosphorylation of the N-terminus threonine and tyrosine residues and (3) dephosphorylation of the T-loop threonine.

We determined that the regulatory schematic presented at the beginning of this chapter, in which cyclin–CDK binding was the only "variable," was too simple to allow for the flexibility needed for a cell to traverse the cell cycle with an appropriate responsiveness to its internal conditions and its surroundings. The model with which we end this chapter has three variables, conferring a level of complexity that allows the CDKs to process a diverse array of environmental signals. Each added variable exponentially increases the number of "inputs" allowed in our regulatory system. We may be tempted to conclude that we now possess an essentially complete understanding of cyclins and CDKs, for surely three variables should allow for enough inputs to serve the cell's needs?

Nature evidently favors complexity, for it has become clear in the past few years that our three–variable model is an incomplete solution. A whole new class of regulatory proteins have emerged, molecules that physically inhibit CDKs, raising the complexity of cell cycle control by yet another exponential factor. These proteins, the CKIs, will be a subject of study in the next chapter.

References

1 Hayles J, Fisher D, Woollard A et al. Temporal order of S phase and mitosis in fission yeast is determined by the state of the p34^{cdc2}–mitotic B cyclin complex. Cell 1994; 78:813–822.
2 Nurse P. Ordering S phase and M phase in the cell cycle. Cell 1994; 79:547–550.
3 Amon A, Tyers M, Futcher B et al. Mechanisms that help the yeast cell cycle clock tick: G2 cyclins transcriptionally activate G2 cyclins and repress G1 cyclins. Cell 1993; 74:993–1007.
4 Amon A, Irniger S, Nasmyth K. Closing the cell cycle circle in yeast: G2 cyclin proteolysis initiated at mitosis persists until the activation of G1 cyclins in the next cycle. Cell 1994; 77:1037–1050.
5 Guadagno TM, Newport JW. Cdk2 kinase is required for entry into mitosis as a positive regulator of Cdc2–cyclin B kinase activity. Cell 1996; 84:73–82.
6 Evans T, Rosenthal ET, Youngblom J et al. Cyclin: a protein specified by maternal mRNA in sea urchin eggs that is destroyed at each cleavage division. Cell 1983; 33:389–396.
7 Murray AW, Kirschner MW. Cyclin synthesis drives the early embryonic cell cycle. Nature 1989; 339:275–280.
8 Wagenaar EB. The timing of synthesis of proteins required for mitosis in the cell cycle of the sea urchin embryo. Exp Cell Res 1983; 144:393–403.
9 Gerhart J, Wu M, Kirschner M. Cell cycle dynamics of an M–phase–specific cytoplasmic factor in Xenopus laevis oocytes and eggs. J Cell Biol 1984; 98:1247–1255.
10 Cyert MS, Kirschner MW. Regulation of MPF activity in vitro. Cell 1988; 53:185–195.
11 Minshull J, Blow JJ, Hunt T. Translation of cyclin mRNA is necessary for extracts of activated Xenopus eggs to enter mitosis. Cell 1989; 56:947–956.
12 Solomon MJ, Glotzer M, Lee TH et al. Cyclin activation of p34^{cdc2}. Cell 1990; 63:1013–1024.
13 Gautier J, Maller JL. Cyclin B in Xenopus oocytes: implications for the mechanism of pre–MPF activation. EMBO J 1991; 10:177–182.
14 Forsburg SL, Nurse P. Cell cycle regulation in the yeasts Saccharomyces cerevisiae and Schizosaccharomyces pombe. Annu Rev Cell Biol 1991; 7:227–256.
15 Nurse P, Thuriaux P, Nasmyth K. Genetic control of the cell division cycle in the fission yeast Schizosaccharomyces pombe. Mol Gen Genet 1976; 146:167–178.

16 Fantes P. Epistatic gene interactions in the control of division in fission yeast. Nature 1979; 279:428–430.
17 Russell P, Nurse P. cdc25+ functions as an inducer in the mitotic control of fission yeast. Cell 1986; 45:145–153.
18 Moreno S, Hayles J, Nurse P. Regulation of $p34^{cdc2}$ protein kinase during mitosis. Cell 1989; 58:361–372.
19 Russell P, Moreno S, Reed SI. Conservation of mitotic controls in fission and budding yeasts. Cell 1989; 57:295–303.
20 Edgar BA, O'Farrell PH. Genetic control of cell division patterns in the Drosophila embryo. Cell 1989; 57:177–187.
21 Jimenez J, Alphey L, Nurse P et al. Complementation of fission yeast $cdc2^{ts}$ and $cdc25^{ts}$ mutants identifies two cell cycle genes from Drosophila: a cdc2 homologue and string. EMBO J 1990; 9:3565–3571.
22 Sadhu K, Reed SI, Richardson H et al. Human homolog of fission yeast cdc25 mitotic inducer is predominantly expressed in G2. Proc Natl Acad Sci USA 1990; 87:5139–5143.
23 Nurse P. Genetic control of cell size at cell division in yeast. Nature 1975; 256:547–551.
24 Nurse P, Thuriaux P. Regulatory genes controlling mitosis in the fission yeast Schizosaccharomyces pombe. Genetics 1980; 96:627–637.
25 Fantes PA. Isolation of cell size mutants of a fission yeast by a new selective method: characterization of mutants and implications for division control mechanisms. J Bacteriol 1981; 146:746–754.
26 Russell P, Nurse P. Negative regulation of mitosis by wee1+, a gene encoding a protein kinase homolog. Cell 1987; 49:559–567.
27 Igarashi M, Nagata A, Jinno S et al. Wee1(+)–like gene in human cells. Nature 1991; 353:80-83.
28 Carr AM, MacNeill SA, Hayles J et al. Molecular cloning and sequence analysis of mutant alleles of the fission yeast cdc2 protein kinase gene: implications for cdc2+ protein structure and function. Mol Gen Genet 1989; 218:41–49.
29 Gould KL, Nurse P. Tyrosine phosphorylation of the fission yeast cdc2+ protein kinase regulates entry into mitosis. Nature 1989; 342:39–45.
30 Gould KL, Moreno S, Owen DJ et al. Phosphorylation at Thr167 is required for Schizosaccharomyces pombe $p34^{cdc2}$ function. EMBO J 1991; 10:3297–3309.
31 Krek W, Nigg EA. Differential phosphorylation of vertebrate $p34^{cdc2}$ kinase at the G1/S and G2/M transitions of the cell cycle: identification of major phosphorylation sites. EMBO J 1991; 10:305–316.
32 Morla AO, Draetta G, Beach D et al. Reversible tyrosine phosphorylation of cdc2: dephosphorylation accompanies activation during entry into mitosis. Cell 1989; 58:193–203.
33 Ducommun B, Brambilla P, Felix MA et al. cdc2 phosphorylation is required for its interaction with cyclin. EMBO J 1991; 10:3311–3319.
34 Gould KL, Moreno S, Tonks NK et al. Complementation of the mitotic activator, $p80^{cdc25}$, by a human protein–tyrosine phosphatase. Science 1990; 250:1573–1576.
35 Dunphy WG, Kumagai A. The cdc25 protein contains an intrinsic phosphatase activity. Cell 1991; 67:189–196.
36 Gautier J, Solomon MJ, Booher RN et al. cdc25 is a specific tyrosine phosphatase that directly activates $p34^{cdc2}$. Cell 1991; 67:197–211.
37 Kumagai A, Dunphy WG. The cdc25 protein controls tyrosine dephosphorylation of the cdc2 protein in a cell–free system. Cell 1991; 64:903–914.
38 Moreno S, Nurse P. Clues to action of cdc25 protein. Nature 1991; 351:194.
39 Strausfeld U, Labbe JC, Fesquet D et al. Dephosphorylation and activation of a $p34^{cdc2}$/cyclin B complex in vitro by human CDC25 protein. Nature 1991; 351:242–245.
40 Lee MS, Ogg S, Xu M et al. cdc25+ encodes a protein phosphatase that dephosphorylates $p34^{cdc2}$. Mol Biol Cell 1992; 3:73–84.
41 Booher RN, Alfa CE, Hyams JS et al. The fission yeast cdc2/cdc13/suc1 protein kinase: regulation of catalytic activity and nuclear localization. Cell 1989; 58:485–497.
42 Ducommun B, Draetta G, Young P et al. Fission yeast cdc25 is a cell–cycle regulated protein. Biochem Biophys Res Commun 1990; 167:301–309.

43 Moreno S, Nurse P, Russell P. Regulation of mitosis by cyclic accumulation of p80^{cdc25} mitotic inducer in fission yeast. Nature 1990; 344:549–552.
44 Featherstone C, Russell P. Fission yeast p107^{wee1} mitotic inhibitor is a tyrosine/serine kinase. Nature 1991; 349:808–811.
45 Parker LL, Atherton–Fessler S, Lee MS et al. Cyclin promotes the tyrosine phosphorylation of p34^{cdc2} in a wee1^{+} dependent manner. EMBO J 1991; 10:1255–1263.
46 Parker LL, Atherton–Fessler S, Piwnica–Worms H. p107^{wee1} is a dual–specificity kinase that phosphorylates p34^{cdc2} on tyrosine 15. Proc Natl Acad Sci USA 1992; 89:2917–2921.
47 Kumagai A, Dunphy WG. Regulation of the cdc25 protein during the cell cycle in Xenopus extracts. Cell 1992; 70:139–151.
48 Russell P, Nurse P. The mitotic inducer nim1^{+} functions in a regulatory network of protein kinase homologs controlling the initiation of mitosis. Cell 1987; 49:569–576.
49 Feilotter H, Nurse P, Young PG. Genetic and molecular analysis of cdr1/nim1 in Schizosaccharomyces pombe. Genetics 1991; 127:309–318.
50 Atherton–Fessler S, Parker LL, Geahlen RL et al. Mechanisms of p34^{cdc2} regulation. Mol Cell Biol 1993; 13:1675–1685.
51 De Bondt HL, Rosenblatt J, Jancarik J et al. Crystal structure of cyclin–dependent kinase 2. Nature 1993; 363:595–602.
52 Jeffrey PD, Russo AA, Polyak K et al. Mechanism of CDK activation revealed by the structure of a cyclinA–CDK2 complex. Nature 1995; 376:313–320.
53 Meijer L, Azzi L, Wang JY. Cyclin B targets p34^{cdc2} for tyrosine phosphorylation. EMBO J 1991; 10:1545–1554.
54 Lundgren K, Walworth N, Booher R et al. mik1 and wee1 cooperate in the inhibitory tyrosine phosphorylation of cdc2. Cell 1991; 64:1111–1122.
55 Booher RN, Deshaies RJ, Kirschner MW. Properties of Saccharomyces cerevisiae wee1 and its differential regulation of p34^{CDC28} in response to G1 and G2 cyclins. EMBO J 1993; 12:3417–3426.
56 Lee MS, Enoch T, Piwnica–Worms H. mik1^{+} encodes a tyrosine kinase that phosphorylates p34^{cdc2} on tyrosine 15. J Biol Chem 1994; 269:30530–30537.
57 Parker LL, Sylvestre PJ, Byrnes MJ et al. Identification of a 95–kDa WEE1–like tyrosine kinase in HeLa cells. Proc Natl Acad Sci USA 1995; 92:9638–9642.
58 Parker LL, Piwnica–Worms H. Inactivation of the p34^{cdc2}–cyclin B complex by the human WEE1 tyrosine kinase. Science 1992; 257:1955–1957.
59 McGowan CH, Russell P. Human Wee1 kinase inhibits cell division by phosphorylating p34^{cdc2} exclusively on Tyr15. EMBO J 1993; 12:75–85.
60 Den Haese GJ, Walworth N, Carr AM et al. The Wee1 protein kinase regulates T14 phosphorylation of fission yeast Cdc2. Mol Biol Cell 1995; 6:371–385.
61 Atherton–Fessler S, Liu F, Gabrielli B et al. Cell cycle regulation of the p34^{cdc2} inhibitory kinases. Mol Biol Cell 1994; 5:989–1001.
62 Kornbluth S, Sebastian B, Hunter T et al. Membrane localization of the kinase which phosphorylates p34^{cdc2} on threonine 14. Mol Biol Cell 1994; 5:273–282.
63 Mueller PR, Coleman TR, Kumagai A et al. Myt1: a membrane–associated inhibitory kinase that phosphorylates Cdc2 on both threonine–14 and tyrosine–15. Science 1995; 270:86–90.
64 Krek W, Nigg EA. Mutations of p34^{cdc2} phosphorylation sites induce premature mitotic events in HeLa cells: evidence for a double block to p34^{cdc2} kinase activation in vertebrates. EMBO J 1991; 10:3331–3341.
65 Norbury C, Blow J, Nurse P. Regulatory phosphorylation of the p34^{cdc2} protein kinase in vertebrates. EMBO J 1991; 10:3321–3329.
66 Pickham KM, Meyer AN, Li J et al. Requirement of mosXe protein kinase for meiotic maturation of Xenopus oocytes induced by a cdc2 mutant lacking regulatory phosphorylation sites. Mol Cell Biol 1992; 12:3192–3203.
67 Smythe C, Newport JW. Coupling of mitosis to the completion of S phase in Xenopus occurs via modulation of the tyrosine kinase that phosphorylates p34^{cdc2}. Cell 1992; 68:787–797.
68 Coleman TR, Tang Z, Dunphy WG. Negative regulation of the wee1 protein kinase by direct action of the nim1/cdr1 mitotic inducer. Cell 1993; 72:919–929.

69 Parker LL, Walter SA, Young PG et al. Phosphorylation and inactivation of the mitotic inhibitor Wee1 by the nim1/cdr1 kinase. Nature 1993; 363:736–738.
70 Wu L, Russell P. Nim1 kinase promotes mitosis by inactivating Wee1 tyrosine kinase. Nature 1993; 363:738–741.
71 Warbrick E, Fantes PA. The wis1 protein kinase is a dosage–dependent regulator of mitosis in Schizosaccharomyces pombe. EMBO J 1991; 10:4291–4299.
72 Warbrick E, Fantes PA. Five novel elements involved in the regulation of mitosis in fission yeast. Mol Gen Genet 1992; 232:440–446.
73 Fantes PA, Warbrick E, Hughes DA et al. New elements in the mitotic control of the fission yeast Schizosaccharomyces pombe. Cold Spring Harb Symp Quant Biol 1991; 56:605–611.
74 Tang Z, Coleman TR, Dunphy WG. Two distinct mechanisms for negative regulation of the Wee1 protein kinase. EMBO J 1993; 12:3427–3436.
75 Honda R, Tanaka H, Ohba Y et al. Mouse $p87^{wee1}$ kinase is regulated by M-phase specific phosphorylation. Chromosome Res 1995; 3:300–308.
76 McGowan CH, Russell P. Cell cycle regulation of human WEE1. EMBO J 1995; 14:2166–2175.
77 Mueller PR, Coleman TR, Dunphy WG. Cell cycle regulation of a Xenopus Wee1–like kinase. Mol Biol Cell 1995; 6:119–134.
78 Watanabe N, Broome M, Hunter T. Regulation of the human WEE1Hu CDK tyrosine 15–kinase during the cell cycle. EMBO J 1995; 14:1878–1891.
79 Ottilie S, Chernoff J, Hannig G et al. The fission yeast genes $pyp1^{+}$ and $pyp2^{+}$ encode protein tyrosine phosphatases that negatively regulate mitosis. Mol Cell Biol 1992; 12:5571–5580.
80 Millar JB, Russell P, Dixon JE et al. Negative regulation of mitosis by two functionally overlapping PTPases in fission yeast. EMBO J 1992; 11:4943–4952.
81 Hannig G, Ottilie S, Erikson RL. Negative regulation of mitosis in fission yeast by catalytically inactive pyp1 and pyp2 mutants. Proc Natl Acad Sci USA 1994; 91:10084–10088.
82 Belle R, Ollivier E, Guerrucci MA. Higher eucaryotic cdc25 proteins are structurally related to phosphoseryl/threonyl protein phosphatases. Biol Cell 1992; 75:139–143.
83 Sebastian B, Kakizuka A, Hunter T. Cdc25M2 activation of cyclin–dependent kinases by dephosphorylation of threonine–14 and tyrosine–15. Proc Natl Acad Sci USA 1993; 90:3521–3524.
84 Millar JB, Blevitt J, Gerace L et al. $p55^{CDC25}$ is a nuclear protein required for the initiation of mitosis in human cells. Proc Natl Acad Sci USA 1991; 88:10500–10504.
85 Galaktionov K, Beach D. Specific activation of cdc25 tyrosine phosphatases by B-type cyclins: evidence for multiple roles of mitotic cyclins. Cell 1991; 67:1181–1194.
86 Hoffmann I, Draetta G, Karsenti E. Activation of the phosphatase activity of human cdc25A by a cdk2–cyclin E dependent phosphorylation at the G1/S transition. EMBO J 1994; 13:4302–4310.
87 Jinno S, Suto K, Nagata A et al. Cdc25A is a novel phosphatase functioning early in the cell cycle. EMBO J 1994; 13:1549–1556.
88 Jessus C, Beach D. Oscillation of MPF is accompanied by periodic association between cdc25 and cdc2–cyclin B. Cell 1992; 68:323–332.
89 Millar JB, Russell P. The cdc25 M–phase inducer: an unconventional protein phosphatase. Cell 1992; 68:407–410.
90 Izumi T, Walker DH, Maller JL. Periodic changes in phosphorylation of the Xenopus cdc25 phosphatase regulate its activity. Mol Biol Cell 1992; 3:927–939.
91 Hoffmann I, Clarke PR, Marcote MJ et al. Phosphorylation and activation of human cdc25-C by cdc2—cyclin B and its involvement in the self–amplification of MPF at mitosis. EMBO J 1993; 12:53–63.
92 Izumi T, Maller JL. Elimination of cdc2 phosphorylation sites in the cdc25 phosphatase blocks initiation of M–phase. Mol Biol Cell 1993; 4:1337–1350.
93 Villa–Moruzzi E. Activation of the cdc25C phosphatase in mitotic HeLa cells. Biochem Biophys Res Commun 1993; 196:1248–1254.
94 Strausfeld U, Fernandez A, Capony JP et al. Activation of $p34^{cdc2}$ protein kinase by microinjection of human cdc25C into mammalian cells. Requirement for prior phosphorylation of cdc25C by $p34^{cdc2}$ on sites phosphorylated at mitosis. J Biol Chem 1994; 269:5989–6000.
95 Hoffmann I, Karsenti E. The role of cdc25 in checkpoints and feedback controls in the eukaryotic cell cycle. J Cell Sci Suppl 1994; 18:75–79.

96 Ogg S, Gabrielli B, Piwnica–Worms H. Purification of a serine kinase that associates with and phosphorylates human Cdc25C on serine 216. J Biol Chem 1994; 269:30461–30469.
97 Izumi T, Maller JL. Phosphorylation and activation of the Xenopus Cdc25 phosphatase in the absence of Cdc2 and Cdk2 kinase activity. Mol Biol Cell 1995; 6:215–226.
98 Galaktionov K, Jessus C, Beach D. Raf1 interaction with Cdc25 phosphatase ties mitogenic signal transduction to cell cycle activation. Genes Dev 1995; 9:1046–1058.
99 Galaktionov K, Chen X, Beach D. Cdc25 cell–cycle phosphatase as a target of c-myc. Nature 1996; 382:511–517.
100 Millar JB, Lenaers G, Russell P. Pyp3 PTPase acts as a mitotic inducer in fission yeast. EMBO J 1992; 11:4933–4941.
101 Mondesert O, Moreno S, Russell P. Low molecular weight protein–tyrosine phosphatases are highly conserved between fission yeast and man. J Biol Chem 1994; 269:27996–27999.
102 Modesti A, Cirri P, Raugei G et al. Expression, purification and kinetic behaviour of fission yeast low M(r) protein–tyrosine phosphatase. FEBS Lett 1995; 375:235–238.
103 Zhang ZY, Zhou G, Denu JM et al. Purification and characterization of the low molecular weight protein tyrosine phosphatase, Stp1, from the fission yeast Schizosaccharomyces pombe. Biochemistry 1995; 34:10560–10568.
104 Picard A, Capony JP, Brautigan DL et al. Involvement of protein phosphatases 1 and 2A in the control of M phase–promoting factor activity in starfish. J Cell Biol 1989; 109:3347–3354.
105 Brautigan DL, Sunwoo J, Labbe JC et al. Cell cycle oscillation of phosphatase inhibitor–2 in rat fibroblasts coincident with $p34^{cdc2}$ restriction. Nature 1990; 344:74–78.
106 Felix MA, Cohen P, Karsenti E. Cdc2 H1 kinase is negatively regulated by a type 2A phosphatase in the Xenopus early embryonic cell cycle: evidence from the effects of okadaic acid. EMBO J 1990; 9:675-683.
107 Yamashita K, Yasuda H, Pines J et al. Okadaic acid, a potent inhibitor of type 1 and type 2A protein phosphatases, activates cdc2/H1 kinase and transiently induces a premature mitosis–like state in BHK21 cells. EMBO J 1990; 9:4331–4338.
108 Devault A, Cavadore JC, Fesquet D et al. Concerted roles of cyclin A, $cdc25^{+}$ mitotic inducer, and type 2A phosphatase in activating the cyclin B/cdc2 protein kinase at the G2/M phase transition. Cold Spring Harb Symp Quant Biol 1991; 56:503–513.
109 Lee TH, Solomon MJ, Mumby MC et al. INH, a negative regulator of MPF, is a form of protein phosphatase 2A. Cell 1991; 64:415–423.
110 Lorca T, Fesquet D, Zindy F et al. An okadaic acid–sensitive phosphatase negatively controls the cyclin degradation pathway in amphibian eggs. Mol Cell Biol 1991; 11:1171–1175.
111 Picard A, Labbe JC, Barakat H et al. Okadaic acid mimics a nuclear component required for cyclin B–cdc2 kinase microinjection to drive starfish oocytes into M phase. J Cell Biol 1991; 115:337–344.
112 Devault A, Fesquet D, Cavadore JC et al. Cyclin A potentiates maturation–promoting factor activation in the early Xenopus embryo via inhibition of the tyrosine kinase that phosphorylates cdc2. J Cell Biol 1992; 118:1109–1120.
113 Fernandez A, Brautigan DL, Lamb NJ. Protein phosphatase type 1 in mammalian cell mitosis: chromosomal localization and involvement in mitotic exit. J Cell Biol 1992; 116:1421–1430.
114 Yanagida M, Kinoshita N, Stone EM et al. Protein phosphatases and cell division cycle control. Ciba Found Symp 1992; 170:130–140, discussion 140 146.
115 Kinoshita N, Yamano H, Niwa H et al. Negative regulation of mitosis by the fission yeast protein phosphatase ppa2. Genes Dev 1993; 7:1059–1071.
116 Dohadwala M, da Cruz e Silva EF, Hall FL et al. Phosphorylation and inactivation of protein phosphatase 1 by cyclin–dependent kinases. Proc Natl Acad Sci USA 1994; 91:6408–6412.
117 Lee TH, Turck C, Kirschner MW. Inhibition of cdc2 activation by INH/PP2A. Mol Biol Cell 1994; 5:323–338.
118 Smith LD, Ecker RE. The interaction of steroids with Rana pipiens oocytes in the induction of maturation. Dev Biol 1971; 25:232–247.
119 Masui Y, Markert CL. Cytoplasmic control of nuclear behavior during meiotic maturation of frog oocytes. J Exp Zool 1971; 177:129–145.

120 Wasserman WJ, Masui Y. Effects of cyclohexamide on a cytoplasmic factor initiating meiotic maturation in Xenopus oocytes. Exp Cell Res 1975; 91:381-388.
121 Dunphy WG. The decision to enter mitosis. Trends Cell Biol 1994; 4:202–207.
122 Amon A, Surana U, Muroff I et al. Regulation of $p34^{CDC28}$ tyrosine phosphorylation is not required for entry into mitosis in S. cerevisiae. Nature 1992; 355:368–371.
123 Sorger PK, Murray AW. S-phase feedback control in budding yeast independent of tyrosine phosphorylation of p34cdc28. Nature 1992; 355:365–368.
124 Yamashita M, Kajiura H, Tanaka T et al. Molecular mechanisms of the activation of maturation–promoting factor during goldfish oocyte maturation. Dev Biol 1995; 168:62–75.
125 Edgar BA, Sprenger F, Duronio RJ et al. Distinct molecular mechanisms regulate cell cycle timing at successive stages of Drosophila embryogenesis. Genes Dev 1994; 8:440–452.
126 Solomon MJ, Lee T, Kirschner MW. Role of phosphorylation in $p34^{cdc2}$ activation: identification of an activating kinase. Mol Biol Cell 1992; 3:13–27.
127 Desai D, Gu Y, Morgan DO. Activation of human cyclin–dependent kinases in vitro. Mol Biol Cell 1992; 3:571–582.
128 Desai D, Wessling HC, Fisher RP et al. Effects of phosphorylation by CAK on cyclin binding by CDC2 and CDK2. Mol Cell Biol 1995; 15:345–350.
129 Morgan DO, De Bondt HL. Protein kinase regulation: insights from crystal structure analysis. Curr Opin Cell Biol 1994; 6:239–246.
130 Russo AA, Jeffrey PD, Pavletich NP. Structural basis of cyclin–dependent kinase activation by phosphorylation. Nature Struct Biol 1996; 3:696–700.
131 Shuttleworth J, Godfrey R, Colman A. $p40^{MO15}$, a cdc2–related protein kinase involved in negative regulation of meiotic maturation of Xenopus oocytes. EMBO J 1990; 9:3233–3240.
132 Fesquet D, Labbe JC, Derancourt J et al. The MO15 gene encodes the catalytic subunit of a protein kinase that activates cdc2 and other cyclin–dependent kinases (CDKs) through phosphorylation of Thr161 and its homologues. EMBO J 1993; 12:3111–3121.
133 Poon RY, Yamashita K, Adamczewski JP et al. The cdc2–related protein $p40^{MO15}$ is the catalytic subunit of a protein kinase that can activate $p33^{cdk2}$ and $p34^{cdc2}$. EMBO J 1993; 12:3123–3132.
134 Solomon MJ, Harper JW, Shuttleworth J. CAK, the $p34^{cdc2}$ activating kinase, contains a protein identical or closely related to $p40^{MO15}$. EMBO J 1993; 12:3133–3142.
135 Darbon JM, Devault A, Taviaux S et al. Cloning, expression and subcellular localization of the human homolog of $p40^{MO15}$ catalytic subunit of cdk–activating kinase. Oncogene 1994; 9:3127–3138.
136 Levedakou EN, He M, Baptist EW et al. Two novel human serine/threonine kinases with homologies to the cell cycle regulating Xenopus MO15, and NIMA kinases: cloning and characterization of their expression pattern. Oncogene 1994; 9:1977–1988.
137 Matsuoka M, Kato JY, Fisher RP et al. Activation of cyclin–dependent kinase 4 (cdk4) by mouse MO15–associated kinase. Mol Cell Biol 1994; 14:7265–7275.
138 Tassan JP, Schultz SJ, Bartek J et al. Cell cycle analysis of the activity, subcellular localization, and subunit composition of human CAK (CDK–activating kinase). J Cell Biol 1994; 127:467–478.
139 Wu L, Yee A, Liu L et al. Molecular cloning of the human CAK1 gene encoding a cyclin–dependent kinase–activating kinase. Oncogene 1994; 9:2089–2096.
140 Fisher RP, Morgan DO. A novel cyclin associates with MO15/CDK7 to form the CDK–activating kinase. Cell 1994; 78:713–724.
141 Makela TP, Tassan JP, Nigg EA et al. A cyclin associated with the CDK–activating kinase MO15. Nature 1994; 371:254–257.
142 Labbe JC, Martinez AM, Fesquet D et al. $p40^{MO15}$ associates with a p36 subunit and requires both nuclear translocation and Thr176 phosphorylation to generate cdk–activating kinase activity in Xenopus oocytes. EMBO J 1994; 13:5155–5164.
143 Stepanova LY, Ershler MA, Belyavsky AV. Sequence of the cDNA encoding murine CRK4 protein kinase. Gene 1994; 149:321–324.
144 Damagnez V, Makela TP, Cottarel G. Schizosaccharomyces pombe Mop1–Mcs2 is related to mammalian CAK. EMBO J 1995; 14:6164–6172.

145 Buck V, Russell P, Millar JB. Identification of a cdk–activating kinase in fission yeast. EMBO J 1995; 14:6173–6183.
146 Poon RY, Yamashita K, Howell M et al. Cell cycle regulation of the $p34^{cdc2}/p33^{cdk2}$–activating kinase $p40^{MO15}$. J Cell Sci 1994; 107:2789–2799.
147 Brown AJ, Jones T, Shuttleworth J. Expression and activity of $p40^{MO15}$, the catalytic subunit of cdk–activating kinase, during Xenopus oogenesis and embryogenesis. Mol Biol Cell 1994; 5:921–932.
148 Gyuris J, Golemis E, Chertkov H et al. Cdi1, a human G1 and S phase protein phosphatase that associates with Cdk2. Cell 1993; 75:791–803.
149 Hannon GJ, Casso D, Beach D. KAP: a dual specificity phosphatase that interacts with cyclin–dependent kinases. Proc Natl Acad Sci USA 1994; 91:1731–1735.
150 Poon RY, Hunter T. Dephosphorylation of Cdk2 Thr160 by the cyclin–dependent kinase–interacting phosphatase KAP in the absence of cyclin. Science 1995; 270:90–93.
151 Fisher RP, Jin P, Chamberlin HM et al. Alternative mechanisms of CAK assembly require an assembly factor or an activating kinase. Cell 1995; 83:47–57.
152 Devault A, Martinez AM, Fesquet D et al. MAT1 ('menage a trois') a new RING finger protein subunit stabilizing cyclin H-cdk7 complexes in starfish and Xenopus CAK. EMBO J 1995; 14:5027–5036.
153 Yee A, Nichols MA, Wu L et al. Molecular cloning of CDK7–associated human MAT1, a cyclin–dependent kinase–activating kinase (CAK) assembly factor. Cancer Res 1995; 55:6058–6062.
154 Adamczewski JP, Rossignol M, Tassan JP et al. MAT1, cdk7 and cyclin H form a kinase complex which is UV light–sensitive upon association with TFIIH. EMBO J 1996; 15:1877–1884.
155 Feaver WJ, Svejstrup JQ, Henry NL et al. Relationship of CDK–activating kinase and RNA polymerase II CTD kinase TFIIH/TFIIK. Cell 1994; 79:1103–1109.
156 Serizawa H, Makela TP, Conaway JW et al. Association of Cdk–activating kinase subunits with transcription factor TFIIH. Nature 1995; 374:280–282.
157 Shiekhattar R, Mermelstein F, Fisher RP et al. Cdk–activating kinase complex is a component of human transcription factor TFIIH. Nature 1995; 374:283–287.
158 Cismowski MJ, Laff GM, Solomon MJ et al. KIN28 encodes a C-terminal domain kinase that controls mRNA transcription in Saccharomyces cerevisiae but lacks cyclin–dependent kinase–activating kinase (CAK) activity. Mol Cell Biol 1995; 15:2983–2992.
159 Kaldis P, Sutton A, Solomon MJ. The CDK–activating kinase (CAK) from budding yeast. Cell 1996; 86:553–564.
160 Thuret JY, Valay JG, Faye G et al. Civ1 (CAK in vivo), a novel Cdk–activating kinase. Cell 1996; 86:565–576.
161 Gabrielli BG, Lee MS, Walker DH et al. Cdc25 regulates the phosphorylation and activity of the Xenopus cdk2 protein kinase complex. J Biol Chem 1992; 267:18040–18046.
162 Gu Y, Rosenblatt J, Morgan DO. Cell cycle regulation of CDK2 activity by phosphorylation of Thr160 and Tyr15. EMBO J 1992; 11:3995–4005
163 Connell–Crowley L, Solomon MJ, Wei N et al. Phosphorylation independent activation of human cyclin–dependent kinase 2 by cyclin A in vitro. Mol Biol Cell 1993; 4:79–92.
164 Kato JY, Matsuoka M, Strom DK et al. Regulation of cyclin D-dependent kinase 4 (cdk4) by cdk4–activating kinase. Mol Cell Biol 1994; 14:2713–2721.

Chapter 5

Cyclin–CDK Regulation: Checkpoints and Inhibitors

Thus far we have developed a working model of vertebrate CDK1 regulation that features three regulatory devices: the association/disassociation of the cyclin and CDK subunits of the maturation–promoting factor (MPF) complex, the phosphorylation vs. dephosphorylation of the Thr14 and Tyr15 residues, and the phosphorylation vs. dephosphorylation of the Thr161 residue. Having elucidated the three elements of this model by the early 1990s, investigators attempted to utilize them in explaining two regulatory puzzles associated with mitosis. First, how does a cell confirm that the replication of its DNA has been concluded by the time it begins mitosis, such that each descendant cell receives a complete and unadulterated genome? Second, how does a cell ensure that any DNA damage that occurred in the duration of the cell cycle has been repaired prior to the onset of DNA replication or mitosis? In seeking the answers to these questions, investigators uncovered several regulatory devices known as checkpoints upstream of cyclin–CDKs and their kinases/phosphatases. In the process, they also serendipitously brought to light a novel class of cyclin–CDK regulators, the physical inhibitors known as CKIs.

Two Regulatory Puzzles

That a more or less faithful copy of a cell's genome should be passed on to the cell's descendants, from one generation to the next, is obvious. Excessive derangements or deletions in the inherited genetic material are likely to impair a progeny cell's function, dooming the greater organism to debilitation. Indeed, the disruption of genes whose products are thought to be key players in the surveillance of DNA damage and the execution of DNA repair appears to be correlated with tumorigenesis, among other disorders.[1-4]

Not surprisingly, cells have developed safeguards to make the initiation of mitosis contingent upon the proper completion of DNA replication. This forestalls the possibility of a daughter cell receiving an incomplete or altered genome because of premature mitosis. Similarly, cells have evolved a complex, multifaceted network through which DNA damage is assayed and an appropriate response is implemented. One alternative is to halt the progression of the cell cycle, facilitate the repair of the DNA, and then resume the cell cycle after the repair is complete. Another option, if the damage is irreparable, is to undergo a course of programmed cell death, variously known as cell "suicide" or apoptosis. In both cases a cell is prevented from making a flawed copy of the genome from a damaged template during DNA replication and potentially bequeathing corrupted genetic material to its descendants.

The mechanisms by which the cell monitors for genome integrity and for the completion of DNA synthesis largely remain a mystery. Investigators invoked the concept of checkpoints when they found that cells arrested at consistent points during the cell cycle in response to DNA–damaging agents and DNA synthesis inhibitors. Checkpoints are best described as biochemical sensing mechanisms that assay the cellular environment; only when the appropriate conditions are fulfilled can a cell proceed past the checkpoints and continue its progression through the cell cycle.[5-7] The checkpoints are thus "sentries" that "watch" specific upstream cellular events and prevent downstream events from occurring prematurely. In our discussion we have already identified several checkpoints, arguably the most important in cell cycle regulation—namely, the late G1 phase restriction (R) point, the G1/S phase transition, and the G2/M phase transition. We should expect there to be a number of checkpoints monitoring crucial events that *must* take place in a proper sequence, e.g., events during cell division such as chromosome condensation, spindle formation, chromosome segregation, and cytokinesis. Accumulated evidence suggests that, in mammalian cells, DNA damage or blockage of replication induced by irradiation or chemical treatment can halt the cell cycle via a number of checkpoints staggered throughout the G1, S and G2 phases.[8-14] In the same vein, various checkpoints dispersed throughout interphase have been identified in both *Saccharomyces cerevisiae* and *Schizosaccharomyces pombe*.[15-20]

At the heart of each of these checkpoints there must lie some control protein or proteins that serve as the focal point of a regulatory pathway. The proteins may be thought of as switches. They are generally in the "off" position, kept that way by conditions in the cellular environment; the cell cannot traverse the checkpoint. When the appropriate signals (e.g., those generated when DNA replication is complete) impinge upon the control proteins, they shift to the "on" position, allowing the cell passage into the next part of the cell cycle. The checkpoints are coordinated and staggered at intervals throughout the phases of the cell cycle. At no time are all the switches in the "on" position; rather, a few switches are turned on in one phase of the cell cycle, they are switched off once the cell has passed into the next phase, another set of switches are turned on, etc. The cell is thereby constrained to proceed through the cell cycle via a single orderly path. In ordinary circumstances, the onset of mitosis must take place a significant length of time after the completion of DNA replication because the former is regulated by a checkpoint control protein (MPF) that cannot be turned on until the cell has passed a prior checkpoint, one that is turned on only after DNA synthesis is finished. In extraordinary circumstances, defective checkpoint mechanisms will allow mitosis to occur immediately after DNA synthesis. The result is an inappropriately rapid, albeit orderly, cell cycle progression, precipitating the uncontrolled growth typical of the cancerous state. In the worst case scenario, checkpoints fail altogether and allow mitosis to begin while DNA synthesis is still in progress, dooming the descendant cells to "mitotic catastrophe"—gross genetic abnormalities that are likely to result in death.

It is of great interest which genes encode checkpoint control proteins, for when they suffer mutations they are potential accomplices in carcinogenesis. For the most part the search for control proteins has focused on the *S. cerevisiae* and *S. pombe* systems. The products of several *S. cerevisiae* genes, including *RAD9* (radiation sensitive), *RAD17*, *RAD24* and *MEC3* (mitosis entry checkpoint) have been implicated as the control proteins of the DNA damage checkpoints in budding yeast, whereas *POL2/DUN2* is crucial to the DNA replication checkpoints; additionally, the *MEC1/SAD3/ESR1*, *TEL1* and *MEC2/SAD1/SPK1/RAD53* products are thought to play roles in both the DNA damage and DNA replication checkpoints (Fig. 5.1).[16,19,21-32] In fission yeast the checkpoints appear to involve a number of *rad⁺* (radiation sensitive) and *hus⁺* (hydroxyurea sensitive) genes[33-41] as well as the recently isolated *chk1⁺* gene (Fig. 5.1).[42-44] For a more detailed description of these genes and their cellular functions, the reader is referred to two recent review articles.[45,46]

Studying mammalian cells, investigators have identified two genes as likely candidates to encode checkpoint control proteins: the *p53* tumor suppressor gene and the *ATM* (ataxia–telangiectasia mutated) gene. They are of exceptional interest because their mutations are known to be associated with a genetic predisposition to cancer. Families with a high incidence of Li–Fraumeni syndrome, characterized by a susceptibility to a variety of tumors throughout childhood and adulthood, have been found to carry germline mutations in *p53*.[47-49] The evidence suggests that the p53 protein is the linchpin of a G1 DNA–damage response pathway that causes cells to arrest at the R point when irradiated.[50-55] The important role of *p53* is highlighted by the finding that it is the most commonly mutated

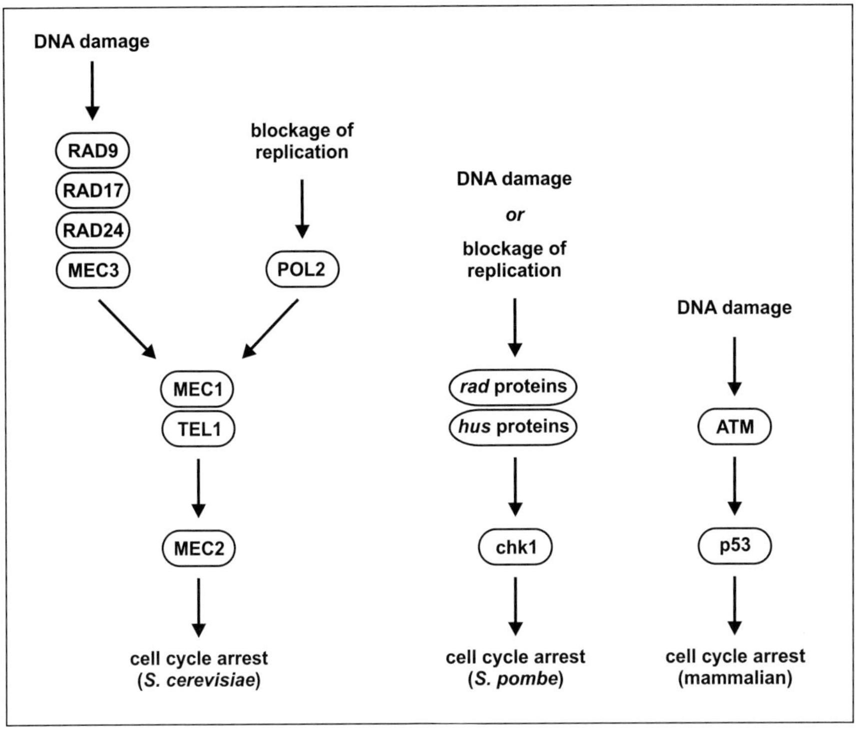

Fig. 5.1. Diagrams depicting putative checkpoint control protein pathways in three systems, *S. cerevisiae*, *S. pombe* and mammalian cells. DNA damage or blockage of DNA replication provokes the activation of specific control proteins, which in turn activate downstream control proteins that lead to a cell cycle arrest, presumably by interfering with cyclin–CDK activity. It should be noted that the pathways portrayed in this figure are simplistic and speculative representations of what actually occurs in cells; it is likely that numerous undiscovered proteins either feed into or diverge from these pathways. The arrows in the figure should be construed as temporal relationships between proteins rather than direct interactions.

gene in human cancers, earning it the name "guardian of the genome."[54] (We will have more to say about p53 in chapter 7.) Human ataxia–telangiectasia (AT), caused by recessive mutations in both copies of the *ATM* gene, is an early onset disease characterized by a high risk of cancer (particularly leukemias and lymphomas), immune system abnormalities, radiation sensitivity, and genetic instability.[56,57] Studies with AT cells suggest that the ATM protein plays a role in G1 and G2 phase DNA–damage response pathways, apparently working in conjunction with the p53 protein.[52,58-61] Interestingly, the human *ATM* gene has been shown to be closely related to the *MEC1* and *TEL1* checkpoint genes in budding yeast.[26,28]

Investigators have thus been able to identify some of the checkpoint control proteins that link the initiation of DNA synthesis to DNA repair and the initiation of mitosis to both DNA repair and the completion of DNA replication (Fig. 5.1), but discovering how these proteins achieve this has proven much more difficult. As more and more evidence lent credence to the cyclin–CDK paradigm of cell cycle regulation, the logical hypothesis that emerged was that the control proteins acted as negative regulators of cyclin–CDK complexes. Having adopted a tripartite model of cyclin–CDK regulation, researchers were intrigued by the possibility that Wee1 and CDC25 were the intermediaries between the control proteins and the CDKs. In other words, the products of such genes as *S. pombe RAD9* and mammalian *p53* might be upstream regulators of the protein kinases and phosphatases that constitute the network that converges upon Wee1 and CDC25 and, through them, cyclin–CDKs.

Attempts to link the yeast checkpoint control pathways with Wee1 and CDC25 have turned up encouraging, if somewhat ambiguous, results. It is clear that in *S. pombe* the requirement of the completion of DNA synthesis for the cell to enter M phase is brokered through the regulation of the balance between the phosphatase and kinase activities of Wee1 and CDC25 respectively. Removal of the activity of Wee1 and its close cousin Mik1 allows the cell to bypass the checkpoint, suggesting that CDK1 is the sole termination point of the checkpoint control pathway.[62] Moreover, several studies suggest that the presence of unreplicated DNA inhibits CDK1 activity by suppressing CDC25; a similar conclusion is supported by experiments in human lymphoma cells.[34,63-65] Using cell–free extracts from *Xenopus laevis* eggs, one group has suggested that incompletely replicated DNA modulates Wee1 activity; in contrast, another group using the same system observed that neither Wee1 nor CDC25 activity is altered in reaction to treatment with a DNA synthesis inhibitor.[66,67]

As for the G2 phase DNA damage checkpoint in fission yeast, studies have implicated Wee1 as the agent through which the checkpoint control pathway obviates CDK1 activity,[68-70] although another study has contradicted this finding.[71] In human keratinocytes, DNA damage appears to induce G2 arrest through the inhibitory phosphorylation of CDK1, presumably the result of a preponderance of Wee1 activity over CDC25 activity.[72]

These findings indicate that Wee1 and CDC25 serve as the intermediates in at least some of the checkpoint control pathways, but there is much left unanswered. As of yet none of these pathways has yielded to a complete molecular dissection. Perhaps the best understood is the *S. pombe* G2 phase DNA damage checkpoint. There is evidence that the *chk1* protein acts downstream of the *rad* proteins and upstream of CDK1.[42,44] But there is no firm comprehension of how *rad* proteins interact with *chk1*, or how *chk1* interacts with CDK1, or how Wee1 and CDC25 might be involved.

Studies of G1 regulation by DNA damage in mammalian cells have identified p53 as a significant modulator of cell cycle progression. Working in mammalian systems, investigators have long sought an understanding of the mechanism(s) by which p53 brings about a late G1 phase cell cycle arrest in response to DNA damage. Armed with the knowledge that p53 is a transcription factor (see chapter 7), the logical place for them to begin a search was among the genes that are transcriptionally activated by p53. In 1993 one collaboration identified through the use of a subtractive hybridization technique a potential candidate, a novel gene dubbed wild–type p53–activated fragment 1 (*WAF1*) that was observed to be highly induced in the presence of p53.[73] They found that *WAF1* was present in a number of tissues and was conserved across species; further, introduction of the gene into various tumor cell lines caused growth suppression similar to that caused by p53. It was thus established that the gene was most likely an important mediator of p53 function. Perhaps the discoverers hoped that the protein product of *WAF1* would prove to be an upstream regulator of either Wee1 or CDC25. It soon became clear, however, that they had something far more interesting on their hands.

The First of the Inhibitors

Serendipitously, the discovery of *WAF1* as a target of p53 transactivation was concomitant with the identification of the encoded protein as a component of cyclin–CDK complexes. It had previously been shown that in untransformed human fibroblasts cyclins A, B and D reside in quaternary complexes along with their respective CDK partners, proliferating cell nuclear antigen (PCNA; a factor that plays a crucial role in DNA replication) and an unidentified 21 kDa protein designated p21.[74,75] Neither PCNA nor p21 was seen to be phosphorylated, suggesting that they were not substrates of the associated cyclin–CDK complexes. Intriguingly, p21 was no longer a component of the complexes in cells transformed with tumor viruses or deficient for p53.[76]

Suspecting that the absence of p21 might be contributory to the deregulation of the cell cycle, the researchers isolated the gene encoding p21 and used the product to reconstruct quaternary complexes with cyclin–CDKs and PCNA, as well as ternary complexes with the cyclin–CDKs alone. In all combinations studied (cyclin A–CDK2, cyclin B–CDK1, cyclin D–CDK4, cyclin E–CDK2) the presence of p21 abrogated kinase activity, albeit with varied efficiencies, implicating the protein as a universal inhibitor of cyclin–CDK complexes.[77]

Concurrently, two other groups independently identified this inhibitor. Using the two–hybrid yeast screening system with a human cDNA library, the one group sought regulators of CDK2 that physically interacted with the kinase.[78] They isolated a gene they named *CIP1* (CDK–interacting protein), from which they obtained protein by means of in vitro translation. The Cip1 protein impaired the kinase activity of CDK2 partnered with various cyclins in vitro; moreover, it was detected in association with various cyclin–CDK combinations in human cells. The other group extracted from mouse fibroblasts protein complexes containing a cyclin, CDK2, and a 20 kDa protein dubbed CAP20.[79] Using biochemical techniques, the investigators separated out the fraction with the CAP20–containing complexes and found that the fraction exhibited minimal kinase activity; this observation suggested that CAP20 acted as a CDK inhibitor in vivo. When purified CAP20 was added in low concentrations to cyclin A- or cyclin E-CDK2 complexes all kinase activity disappeared.

In a remarkable convergence of the work of the four teams, the sequences of *WAF1*, *CIP1* and the gene encoding p21 were found to be identical, and CAP20 was shown to have significant amino acid homology to p21.[73,77-79] The 4–fold discovery of the protein marked a milestone in the progress of the cell cycle field, for it provided a solution to a long–standing mystery and revealed the existence of a new class of cell cycle regulators. The characteristics of the p21 protein present an elegant explanation of p53–mediated growth suppression. p53 levels surge in response to DNA damage (this phenomenon is an unresolved biochemical mystery; readers interested in the latest clues are referred to a recent review[80]), and as a transcriptional activator p53 induces the expression of p21. As the quantity of p21 in the cell increases, more and more of the inhibitor molecules bind to cyclin D- and cyclin E-CDK complexes. Unable to phosphorylate their downstream targets, the cyclin–CDKs cannot relieve the cell of late G1 phase checkpoint controls, particularly the R point. The cell arrests in late G1 phase. Moreover, p21's ability to bind PCNA along with the cyclin–CDKs provides a second means by which to inhibit DNA replication. The upshot is that the cell's DNA repair mechanisms are afforded the opportunity to restore the integrity of the genetic material before it is duplicated. Once the repair is complete p53 is restored to its original level (exactly how this might occur is still unclear) and the induction of p21 ceases. As p21 degradation outstrips its synthesis, the cyclin–CDK complexes (as well as PCNA) are freed from p21–enforced inhibition; their restored kinase activity propels the cell forward through the late G1 checkpoints and into subsequent stages of the cell cycle.

Around the time of the discovery of p21, experiments were undertaken to isolate DNA synthesis inhibitors induced in senescent human fibroblasts.[81] One promising candidate, dubbed *SDI1* (senescent cell–derived inhibitor), was found to have 10– to 20–fold increased expression in senescent cells as compared to

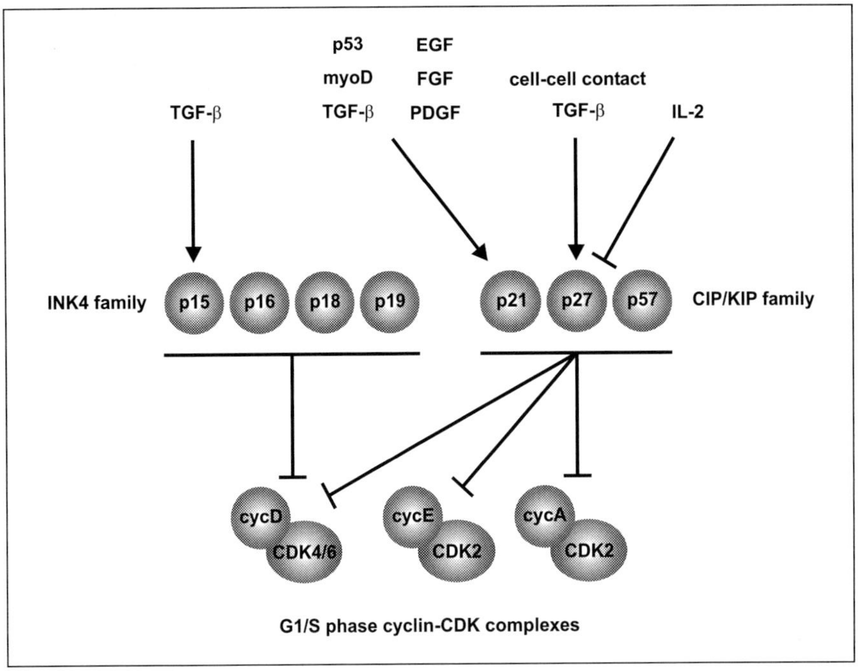

Fig. 5.2. Diagram depicting the cellular roles of the cyclin–CDK inhibitors (CKIs). Arrows indicate an activation or upregulation of the target protein, and bars indicate an inhibition or downregulation of the target protein. Various agonists and stimuli exert positive (or negative) influences on individual CKIs, which in turn inhibit one or more types of G1/S phase cyclin–CDK complexes and thereby induce a cell cycle arrest. The INK4 family of CKIs preferentially inhibit cyclin D–CDK4 and –CDK6 complexes, whereas the CIP/KIP family of CKIs act upon cyclin–CDK complexes containing cyclins D, E or A.

young cells; further, the *SDI1* mRNA level increased in parallel to the onset of the senescent phenotype and loss of cell proliferation. Upon the sequencing of *SDI1* it became clear that the encoded protein, which evidently played a role in cellular aging processes, was none other than a fifth incarnation of the p21 protein.

In the literature one will see the p21 gene variously referred to as *CIP1*, *WAF1*, *CAP20* or *SDI1*, and the protein as Cip1, Waf1, Cap20 or Sdi1. For the sake of simplicity in our nomenclature, we will refer to the protein as p21 and the gene as *p21*, and the same convention will be applied to other similar proteins.

The p21 protein has proven to be only the first of an emerging class of CDK regulators, appropriately called CDK inhibitors (CKIs), that operate through physical inhibition (Fig. 5.2). p21 and two partially homologous inhibitors, p27 and p57, constitute the CIP/KIP family of CKIs. The so–called INK4 (inhibitor of CDK4) proteins, p15, p16, p18 and p19, have also been characterized and compose a distinct, second CKI family. (It should be noted that each of these proteins received its name from its approximate molecular weight in kDa.) In the next few sections we will give a detailed treatment of each member of these inhibitor families.

p21

The model in which p53 utilizes p21 to exert its growth–suppressive effects has received support from the observation that wild–type human fibroblasts arrested in G1 phase by γ–irradiation display increased p21 levels, whereas p53–deficient cells exposed to radiation neither arrest in G1 phase nor show an increase in p21.[82] Complementary experimentation has shown that p21 is induced in response to DNA–damaging agents that arrest cell growth via p53, but is not induced by agents that act in a p53–independent manner.[83] Further, researchers creating and using two tumor cell lines differing only in their p21 status have shown that while the p53–mediated DNA damage response was normal in the cells with wild–type p21, the response was abrogated in the p21–deficient cells.[84]

Also lending credence to the model are a number of studies demonstrating that p21 is a dual–specificity inhibitor and, accordingly, has two distinct functional domains. The first, residing in the N-terminal portion of the molecule, mediates the CDK–inhibitory effects; the second, located near the C-terminus of the protein, enables p21 to bind PCNA and inhibit its function, directly blocking DNA replication.[85-93] Perhaps the most convincing evidence emerges from a mouse model in which both *p21* alleles have been replaced with defective alleles by means of homologous recombination techniques employed on embryonic stem cells (such models are colloquially known as "knockout mice"). *p21*–nullizygous mice are viable and to all appearances healthy, but their cells are impaired in their ability to arrest in G1 phase in response to DNA damage.[94,95] It should be noted that the G1 arrest caused by DNA damage is not completely abrogated in p21–minus cells; perhaps not surprisingly this implies other p53 targets or p53–independent pathways also contribute to this arrest.

p21 is a versatile, although not universal, inhibitor of CDKs; it appears to be effective against CDK2, CDK3, CDK4 and CDK6, less effective against CDK1 and CDK5, and ineffective against CDK7.[96] Such is p21's potency that when overexpressed in cells it can single–handedly bring about a cell cycle block; notably, p21 is observed to induce a G1 phase arrest, suggesting that in vivo the inhibitor preferentially acts upon complexes with the G1 cyclins, D and E.[96] The mechanism by which p21 abrogates cyclin–CDK kinase activity has not yet been elucidated (but see p27 discussion below). There is evidence that by virtue of its binding to the cyclin–CDK complex p21 prevents the phosphorylation of the CDK by CAK, but this is a partial answer at best because p21 can bind and inhibit complexes that already have been subjected to CAK phosphorylation.[97,98] Complicating the already confused issue is the finding that p21–containing cyclin–CDK complexes can exist in both catalytically active and inactive forms,[99] coupled with the observation that more than one p21 molecule is needed to bind to the cyclin–CDK in order to ensure that the complex is inactive.[96] It is conceivable that the presence of just one p21 molecule in the complex enhances kinase activity by stabilizing the complex, whereas additional p21 molecules will abrogate kinase activity.

Any protein found to inhibit kinases crucial for cell cycle progression and to be capable of arresting cell growth qualifies as a potential tumor suppressor protein.[100] One commonly–used litmus test for such proteins is a search for mutations or deletions of the suspected protein in a variety of primary tumors; for a bona fide tumor suppressor, defects or deficiencies of the protein would be expected to be highly correlated with the incidence of at least several tumor types. Because of the negative findings of the tumor studies conducted thus far, the verdict for p21 rules against its being a tumor suppressor protein, as we shall see in chapter 8. This does not absolutely rule out a tumor–suppressive function for p21, however. If a related protein or proteins shared in this function, derangement of p21 alone would be insufficient to contribute to the development of cancer. All the related proteins would have to suffer mutations, and the chance of several events is ordinarily much less than that of a single event. The appropriate combination of mutations would probably not appear in tumors with any consistency. An alternative explanation may lie in the hypothesis that a single p21 molecule stabilizes and enhances the activity of a cyclin–CDK complex. Assuming this is true, mutation of p21 might actually impair cyclin–CDK function in certain situations and thus be undesirable in cancer cells, making it a rare contributor to oncogenesis.

Roles for p21 that extend beyond p53–enacted cell cycle control might also influence p21's potential as a tumor suppressor. Elevated p21 levels are observed in senescent cells, linking the inhibitor to the cell's aging processes.[101] A number of studies implicate p21 as a key player in the terminal differentiation of proliferating skeletal muscle cells; the muscle–specific transcription factor MyoD causes a sustained elevation in the expression level of p21 and thereby induces an irreversible cell cycle exit.[102-105] p21 can be upregulated in a p53–independent manner by a number of agonists, including transforming growth factor–beta (TGF–β), platelet–derived growth factor (PDGF), fibroblast growth factor (FGF) and epidermal growth factor (EGF), as well as the intracellular signal transduction protein STAT1, for the purpose of inhibiting the progression of the cell cycle.[106-113] p21 may therefore serve as an intermediary link between the core cell cycle machinery and a variety of upstream signal transduction pathways operating in proliferation control and differentiation.

Intriguingly, p21's inhibitory activity appears not to be limited to cyclin–CDK complexes; a recent study demonstrates that p21 can bind and inhibit the stress–activated protein kinases (SAPKs), a subfamily of the mitogen–activated protein (MAP) kinases, in vitro.[114] The SAPKs are not directly involved in cell cycle regulation but instead in the cell's responses to a variety of stresses such as DNA damage or heat shock. Thus p21 might be a regulator of stress–activated signaling pathways as well as cell proliferation.

p27

p27, also known in the literature as Kip1, was originally characterized as a mediator of extracellular antiproliferative signals, particularly those initiated by treatment with TGF–β or cAMP or by physical contact with neighboring cells.[97,115-118] It was also identified as an inhibitor of cell proliferation that could be inactivated by mitogenic factors such as interleukin-2 (IL–2).[119,120] p27 was of immediate interest to investigators as a cellular target of both positive and negative growth factors. Upon the cloning of the gene, the p27 protein was found to be a potent inhibitor of cyclin–CDK complexes with cyclins D, E and A; moreover, the inhibition was enacted via direct physical interaction with the complexes, reminiscent of p21.[116,121] Indeed, p27 shares homology with p21 in the N–terminal portion of the molecule, the region which in p21 mediates CDK–binding and inhibition. For this reason p21 and p27 are acknowledged as members of the same CKI family. Notably, the two molecules diverge with respect to their C-terminal domains, consistent with the observation that p27 is unable to bind PCNA,[89] although both proteins possess a putative nuclear localization signal in the region. Whether p27 retains some function distinct from p21 by virtue of its C–terminus is as of yet unclear.

Similar to p21, p27 induces a G1 phase arrest when it is overexpressed in cells.[116,121] Consistent with this finding is the observation that fibroblasts deprived of exposure to mitogens accumulate p27; correspondingly, the cells are halted in G1 phase, and the cyclin–CDK complexes responsible for driving the cells past the R point are found in inactive form. Further, these fibroblasts fail to arrest in response to mitogen withdrawal when p27 expression is inhibited with antisense oligonucleotides.[122] Other studies suggest that p27 can contribute to differentiation in some cell lines[123,124] and can counter the malignant phenotype of human brain tumor cells.[125] In sum, the various antiproliferative effects exhibited by p27 implicate it as a tumor suppressor protein. Yet at the present time there is little evidence linking p27 mutations to human tumors of any type. Perhaps p27 shares sufficiently in p21's functional role that both must be mutated (which would be an exceedingly rare occurrence) in order for either to contribute to oncogenesis.

Experiments in genetically altered mice support an important role for p27 in tumor formation and proliferation control in general. *p27*–knockout mice are viable and, interestingly, display enhanced growth; their tissues are enlarged and contain more cells than those of their wild–type counterparts, particularly in the thymus, pituitary, and adrenal glands and gonadal organs.[126-128] There are other manifestations of p27 deficiency, such as pituitary tumorigenesis, dysplasia in the retina, and female sterility, but on the whole the knockout mice reflect an increase in cell proliferation. p27 evidently plays an inhibitory role in growth regulation during development. Further, the contrast in the phenotypes of *p21*– and *p27*–knockout mice suggests that p27 is a participant in the cell cycle beyond being simply a redundant version of p21. It is noteworthy that p27 activity appears to be at its highest in quiescent cells, i.e., cells that have temporarily exited the cell cycle,[97,116,118,120,129] whereas p21 is abundant in senescent cells that have permanently ceased proliferation or in cells that have suffered DNA damage and induced p53 expression.

The means by which p27 inhibits cyclin–CDK complexes has recently been elucidated by the determination of the crystal structure of the N-terminal inhibitory domain of p27 bound to a cyclin A-CDK2 complex.[130] One portion of p27 binds to the cyclin A subunit by means of a Leu–Phe–Gly motif conserved among all Cip/Kip family members. Another portion of the CKI intrudes into the active site of the CDK subunit, making ATP–binding impossible. These two major interactions between p27 and the cyclin–CDK firmly anchor the inhibitor to the complex and preclude the phosphorylation of downstream targets by the CDK's catalytic domain. Presumably this elegant mechanism is applicable to the other Cip/Kip proteins (Fig. 5.3), although there are difficulties in explaining why more than one p21 molecule is needed to inhibit the activity of a cyclin–CDK complex (see previous section). It is conceivable that in some situations p21 cannot readily bind both the cyclin and CDK subunits at once, and that two p21 proteins are needed to ensure that both of the necessary interactions are fulfilled.

The induction of p27 activity by the stimuli described above is (unlike p21) due predominantly to post–translational mechanisms. Although p27 messenger RNA is often present in cells, p27–inducing signals can lead to the stabilization of the protein, or its liberation from a "sequestered" state. The latter mechanism for p27 control will be further explored in chapter 7's discussion of transforming growth factor–beta (TGF–β) function; the former control mechanism returns us once again to the theme of cell cycle control by periodic (or episodic) synthesis and destruction of regulatory machinery. Significantly, p27 has been revealed to be susceptible to ubiquitin–mediated proteolysis.[131] This observation suggests

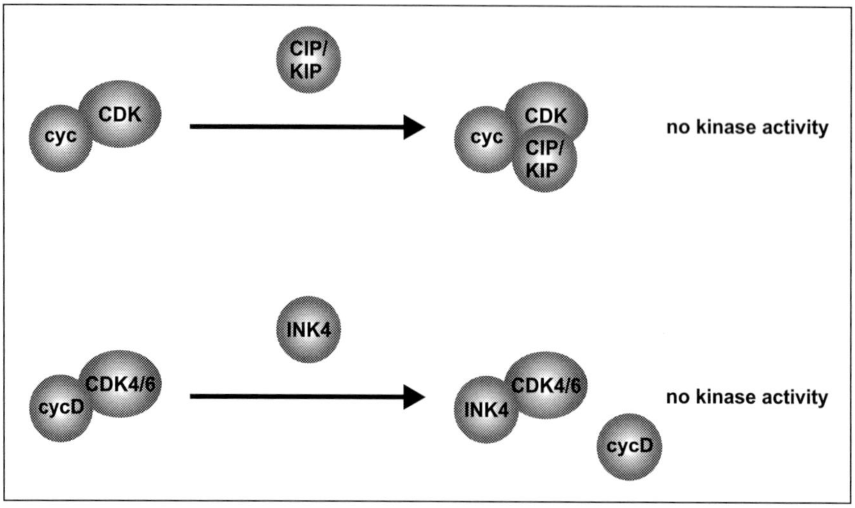

Fig. 5.3. Diagrams depicting the putative mechanisms of cyclin–CDK inhibition by CKIs. CIP/KIP inhibitors are thought to bind to cyclin–CDKs without disrupting the complexes; they abrogate kinase activity by preventing substrates and ATP from coming into contact in the complexes' active sites. INK4 inhibitors (which are specific for cyclin D-CDK complexes) are thought to displace the cyclin subunits from the complexes; in the absence of cyclins, CDKs are unable to phosphorylate substrates.

that the timed destruction of a CKI may be an important mechanism for generating cyclin–CDK activity.[131,132] This contrasts with the ubiquitin–mediated degradation of cyclins A and B in early M phase, which serves to abrogate cyclin–CDK activity (see chapter 3). One can easily imagine a positive growth factor spurring into action the cell's ubiquitinating machinery in late G1 phase, for example, and triggering the destruction of such CKIs as p27 and p16, thereby allowing for the assumption of kinase activity by cyclin D– and cyclin E–CDK complexes and the consequent passage of the cell through the G1/S phase transition. Whether ubiquitin is relevant to the lifespans of CKIs other than p27 remains to be seen.

p57

p57, also known as Kip2, is closely related to p21 and p27.[133,134] More than twice the size of either of its relatives, p57 is composed of four structural domains. Near the N–terminus is a region highly homologous to the CDK–inhibitory N–terminal domains of p21 and p27, and p57's C-terminal region corresponds to the C-terminal domain of p27. Sandwiched between the two terminal domains are a proline–rich domain and a highly acidic domain, both of whose functions are unknown. p57 appears to be a conserved copy of p27 with extra material inserted into its midst.

Functionally p57 very much resembles its two cousins; the protein is a potent inhibitor of the cyclin–CDK complexes present in the G1 and S phases (cyclin D–CDK4 and – CDK6, cyclin E–CDK2, cyclin A–CDK2), and overexpression of p57 causes cells to arrest in G1 phase.[133,134] p57 is a plausible candidate for a tumor suppressor protein, although the evidence is inconclusive (see chapter 8). One basis by which we can distinguish p57 from its two relatives is its expression patterns. Whereas p21 and p27 are ubiquitously expressed in human tissues, p57 is predominantly expressed in highly differentiated tissues such as skeletal muscle, brain, heart and placenta.[133,134] This finding suggests that p57 may be a specialized CKI with an exclusive role in developmental processes.

p16

The first hint of the existence of a group of proteins that preferentially bind to CDK4 and CDK6 was seen in the form of an unidentified 16 kDa protein that associated with CDK4 in human fibroblasts transformed with simian virus 40 (SV40).[76] Shortly thereafter the gene encoding this protein, designated p16, was identified in a two–hybrid interaction screen searching for proteins that could associate with CDK4;[135] the distinctive feature of the predicted amino acid sequence was that it consisted of four tandemly repeated ankyrin motifs.[136] Not only was the p16 protein found to specifically bind CDK4 and CDK6 in vitro and in vivo, it was observed to strongly inhibit cyclin D–CDK4 and –CDK6 complexes.[135,137] The precise mechanism by which p16 (and the other INK4 inhibitors) counters cyclin–CDK activity is still not known, although it is thought that p16 disrupts cyclin–CDK binding through its own binding to the CDK (Fig. 5.3).[135,138,139]

Known in the literature by a number of names including INK4a, MTS1, CDKN2 and CDK4I, the p16 protein is regarded as a strong candidate to be a tumor suppressor. Indeed, *p16* was independently cloned as a gene involved in familial melanoma and deleted in many cultured tumor cell lines, and was thus dubbed *MTS1* (multiple tumor suppressor).[140] Consistent with a tumor suppressor role for p16, ectopic expression of p16 in various cell lines instigates a G1 phase arrest,[141-143] and the introduction of the inhibitor into cancerous cells lacking endogenously–produced p16 suppresses their growth.[144-147] p16 expression has been found to be impaired in such a wide variety of tumor cell lines and primary tumor samples, whether through the deletion or mutation of *p16* or the methylation of nucleotides in the promoter of the wild–type gene, that it is conceivable that p16 is the most commonly altered cell cycle regulator in cancer. A greatly expanded discussion of p16's role in oncogenesis will be offered in chapter 8.

Despite p16's discovery more than three years ago, the nature of its participation in cell cycle regulation remains mysterious. At the present time there are no firmly established upstream inducers of the p16 gene, although elevated p16 levels have been observed in cells entering a senescent state.[148,149] *p16*–knockout mice are viable and fertile, suggesting that p16 is not a crucial participant in developmental processes, although there are phenotypic abnormalities that suggest that p16 is involved in the differentiation of melanocytes and certain hematopoietic lineages. The mice display an increased susceptibility to the development of tumors, both those that arise spontaneously and those that are induced by exposure to carcinogens. Moreover, primary fibroblasts taken from the knockout mice proliferate aberrantly, apparently immortal from the time of plating, and are unusually amenable to neoplastic transformation, consistent with an inability to undergo senescence. These observations lend further support to the speculation that p16 is a bona fide tumor suppressor protein.

Other INK4 Proteins

p15, also known as INK4b or MTS2, was originally identified in TGF–β–treated cells as a protein that bound CDK4 and CDK6.[137] p15 weakly cross–reacted with p16 antiserum, leading to suspicion that p15 and p16 were closely related molecules. Upon sequencing the *p15* gene, investigators discovered that the encoded protein indeed has significant homology with p16, as well as containing the structural pattern of four ankyrin repeats characteristic of p16.[137] Not surprisingly, p15 preferentially inhibits cyclin D–CDK4 and –CDK6 complexes and ignores CDK1, CDK2 and CDK5. As with p16, it appears that p15 eliminates kinase activity by competing with the cyclin subunit for binding to the CDK.[137]

In a number of cell lines exposure to TGF–β induces p15 expression, which in turn precipitates a cell cycle arrest.[137,150] In light of this finding, we should view p15 as a potential tumor suppressor protein. Confusing the issue is the fact that the human *p15* gene is adjacent to the *p16* gene on chromosome 9.[137,151] They are frequently found to be codeleted in a range of tumor types; it is therefore uncertain whether *p15* deletion in and of itself is contributory to oncogenesis or whether it is a trivial phenomenon caused by the proximity of the gene to *p16*. However, the appearance of tumors in *p16*–knockout mice, where the p15 gene remains intact, suggests that loss of *p15* is not obligatory in tumorigenesis, and as will be discussed in chapter 8, loss of p16 function through missense mutation seems to be contributory to cancer even in the absence of alterations in p15.

p18 and p19, also known in the literature as INK4c and INK4d respectively, are homologous to p15 and p16; each contains the four ankyrin repeats characteristic of the INK4 family.[141,152-154] Both proteins specifically inhibit cyclin D-CDK4 and –CDK6 complexes while sparing cyclin E–CDK2, cyclin A–CDK2 and cyclin B–CDK1. Moreover, like p15 and p16, both proteins preferentially form binary complexes with CDK4 and CDK6 in vivo, suggesting that they abrogate cyclin–CDK activity by competitively binding to the CDK subunit and excluding the cyclin subunit from the complex. Not surprisingly, p18 and p19 both display growth–suppressive properties when introduced into cells, although neither appears to be mutated to any great extent in tumors.

What need does the cell have for four proteins that appear to be virtually identical with respect to structure and function? The expression patterns of p18 and p19 in human tissues suggest a partial answer to this question. *p18* mRNA is most highly expressed in skeletal muscle, moderately expressed in pancreas and heart, and virtually absent in placenta and brain.[141] In contrast, *p19* mRNA is most highly expressed in thymus, blood and brain, moderately expressed in spleen, testis, skeletal muscle, and heart, and absent in prostate, ovary, small intestine, colon, liver, kidney and pancreas.[154,155] It is therefore possible that p18 and p19 play specific roles in the development and differentiation of distinct subsets of tissues rather than being involved with the general regulation of the cell cycle.

Other CDK–Interacting Proteins

Our attention has focused on the CKIs as proteins that associate with cyclin–CDK complexes and in doing so inhibit their kinase activity. To balance our discussion, we should consider the theoretical possibility of molecules that can bind cyclin–CDKs and enhance their activity. Such molecules might act as assembly factors, as scaffolding proteins upon which cyclin–CDK complexes would be constructed. Alternatively, they might serve as a "glue" that would stabilize complexes and prevent them from breaking apart before the appropriate time.

To date there is only one promising lead in the search for CDK enhancers. CDC37, whose gene was originally identified in *Saccharomyces cerevisiae* because its mutation caused a cell cycle arrest at the START checkpoint,[156] appears to be required for the association of CDK1 with either G1 cyclins or mitotic cyclins in budding yeast.[157] Moreover, it has recently been reported that the mammalian version of CDC37 binds and stabilizes CDK4, suggesting a role in the production of cyclin D–CDK4 complexes.[158,159]

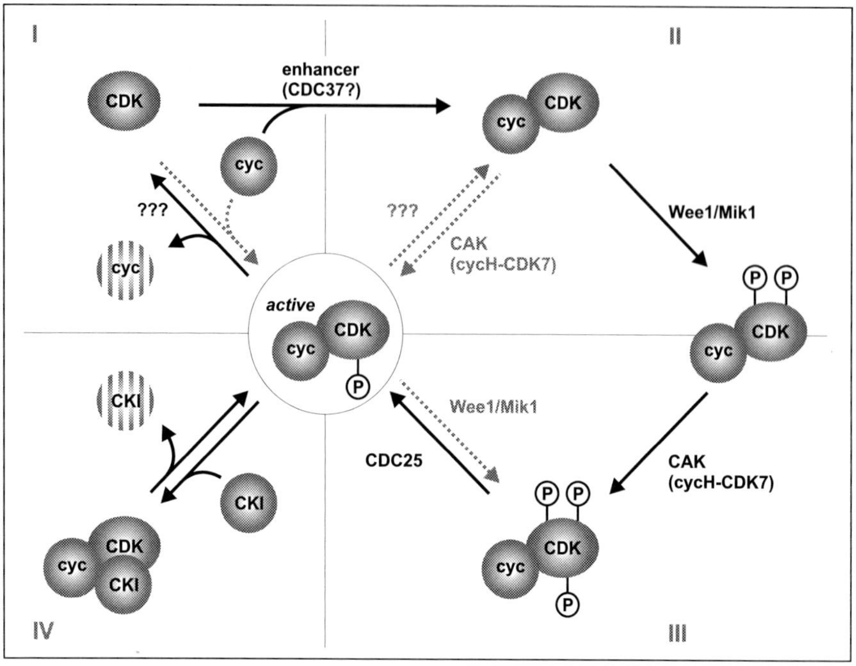

Fig. 5.4. Schematic illustration of the model of activation and inactivation of cyclin–CDK complexes. Each of the four "quadrants" depict one mode of cyclin–CDK regulation: (I) initial formation of the cyclin–CDK complex by the association of the two proteins, possibly facilitated by "enhancer" proteins, and the dissociation of the complex via the destruction of the cyclin; (II) phosphorylation of a threonine residue (Thr161 in CDK1) by CAK, and dephosphorylation of the same residue by an unidentified phosphatase or phosphatases; (III) dephosphorylation of two N-terminal residues (Thr14 and Tyr15 in CDK1) by CDC25, and phosphorylation of the residues by Wee1 and Mik1; (IV) inhibition of the complex by CKIs. The bold arrows trace the sequence of events thought to occur in the cell, whereas the gray dotted arrows represent theoretical events that are not believed to occur.

Although support is lacking for the conjecture, one might guess that proteins such as CDC37 could compete with CKIs for binding to cyclin–CDKs, thereby acting as "anti–CKIs;" perhaps they are produced by cells in response to specific growth–promoting signals. While our understanding of the biochemical regulation and function of the isolated cyclin–CDK complex has grown rapidly, the true nature of CDK activity in intact cells largely remains mysterious. Future work aimed at uncovering all the various interactors impinging on CDK activity, either positively or negatively, will surely reveal an additional complexity to be expected from this central regulatory system.

A Comprehensive Regulatory Model

We have developed a satisfactory albeit elaborate (and possibly incomplete) model of cyclin–CDK regulation (Fig. 5.4). The versatility of the cyclin–CDK complex allows it to accept inputs from a panoply of upstream sources, whether checkpoint controls or signals originating from positive and negative growth factors. Now that we possess a comprehensive understanding of the primary cell cycle regulators—cyclins, CDKs and CKIs—it is time to consider the consequences of cyclin–CDK activation, a task we will take up in the next chapter.

References

1 Enoch T, Norbury C. Cellular responses to DNA damage: cell–cycle checkpoints, apoptosis and the roles of p53 and ATM. Trends Biochem Sci 1995; 20:426–430.

2 Elledge RM, Lee WH. Life and death by p53. Bioessays 1995; 17:923–930.

3 Jackson SP. Cancer predisposition. Ataxia–telangiectasia at the crossroads. Curr Biol 1995; 5:1210–1212.

4 Meyn MS. Ataxia–telangiectasia and cellular responses to DNA damage. Cancer Res 1995; 55:5991–6001.

5 Hartwell LH, Weinert TA. Checkpoints: controls that ensure the order of cell cycle events. Science 1989; 246:629–634.

6 Murray AW. Creative blocks: cell–cycle checkpoints and feedback controls. Nature 1992; 359:599–604.

7 Murray A. Cell cycle checkpoints. Curr Opin Cell Biol 1994; 6:872–876.

8 Little JB. Delayed initiation of DNA synthesis in irradiated human diploid cells. Nature 1968; 218:1064–1065.

9 Busse PM, Bose SK, Jones RW et al. The action of caffeine on X–irradiated HeLa cells. III. Enhancement of X-ray–induced killing during G2 arrest. Radiat Res 1978; 76:292–307.

10 Painter RB, Young BR. Radiosensitivity in ataxia–telangiectasia: a new explanation. Proc Natl Acad Sci USA 1980; 77:7315–7317.

11 Lau CC, Pardee AB. Mechanism by which caffeine potentiates lethality of nitrogen mustard. Proc Natl Acad Sci USA 1982; 79:2942–2946.

12 Painter RB. Inhibition of mammalian cell DNA synthesis by ionizing radiation. Int J Radiat Biol Relat Stud Phys Chem Med 1986; 49:771–781.

13 Nagasawa H, Kraemer KH, Shiloh Y et al. Detection of ataxia telangiectasia heterozygous cell lines by postirradiation cumulative labeling index: measurements with coded samples. Cancer Res 1987; 47:398–402.

14 Schlegel R, Pardee AB. Periodic mitotic events induced in the absence of DNA replication. Proc Natl Acad Sci USA 1987; 84:9025–9029.

15 Wintersberger U, Karwan A. Retardation of cell cycle progression in yeast cells recovering from DNA damage: a study at the single cell level. Mol Gen Genet 1987; 207:320–327.

16 Weinert TA, Hartwell LH. The RAD9 gene controls the cell cycle response to DNA damage in Saccharomyces cerevisiae. Science 1988; 241:317–322.

17 Weinert TA. Dual cell cycle checkpoints sensitive to chromosome replication and DNA damage in the budding yeast Saccharomyces cerevisiae. Radiat Res 1992; 132:141–143.

18 Sheldrick KS, Carr AM. Feedback controls and G2 checkpoints: fission yeast as a model system. Bioessays 1993; 15:775–782.

19 Siede W, Friedberg AS, Friedberg EC. RAD9–dependent G1 arrest defines a second checkpoint for damaged DNA in the cell cycle of Saccharomyces cerevisiae. Proc Natl Acad Sci USA 1993; 90:7985–7989.

20 D'Urso G, Nurse P. Checkpoints in the cell cycle of fission yeast. Curr Opin Genet Dev 1995; 5:12–16.
21 Stern DF, Zheng P, Beidler DR et al. Spk1, a new kinase from Saccharomyces cerevisiae, phosphorylates proteins on serine, threonine, and tyrosine. Mol Cell Biol 1991; 11:987–1001.
22 Weinert TA, Hartwell LH. Cell cycle arrest of cdc mutants and specificity of the RAD9 checkpoint. Genetics 1993; 134:63–80.
23 Allen JB, Zhou Z, Siede W et al. The SAD1/RAD53 protein kinase controls multiple checkpoints and DNA damage–induced transcription in yeast. Genes Dev 1994; 8:2401–2415.
24 Kato R, Ogawa H. An essential gene, ESR1, is required for mitotic cell growth, DNA repair and meiotic recombination in Saccharomyces cerevisiae. Nucleic Acids Res 1994; 22:3104–3112.
25 Weinert TA, Kiser GL, Hartwell LH. Mitotic checkpoint genes in budding yeast and the dependence of mitosis on DNA replication and repair. Genes Dev 1994; 8:652–665.
26 Greenwell PW, Kronmal SL, Porter SE et al. TEL1, a gene involved in controlling telomere length in S. cerevisiae, is homologous to the human ataxia telangiectasia gene. Cell 1995; 82:823–829.
27 Lydall D, Weinert T. Yeast checkpoint genes in DNA damage processing: implications for repair and arrest. Science 1995; 270:1488–1491.
28 Morrow DM, Tagle DA, Shiloh Y et al. TEL1, an S. cerevisiae homolog of the human gene mutated in ataxia telangiectasia, is functionally related to the yeast checkpoint gene MEC1. Cell 1995; 82:831–840.
29 Navas TA, Zhou Z, Elledge SJ. DNA polymerase epsilon links the DNA replication machinery to the S phase checkpoint. Cell 1995; 80:29–39.
30 Paulovich AG, Hartwell LH. A checkpoint regulates the rate of progression through S phase in S. cerevisiae in response to DNA damage. Cell 1995; 82:841–847.
31 Sanchez Y, Desany BA, Jones WJ et al. Regulation of RAD53 by the ATM–like kinases MEC1 and TEL1 in yeast cell cycle checkpoint pathways. Science 1996; 271:357–360.
32 Sun Z, Fay DS, Marini F et al. Spk1/Rad53 is regulated by Mec1–dependent protein phosphorylation in DNA replication and damage checkpoint pathways. Genes Dev 1996; 10:395–406.
33 Murray JM, Carr AM, Lehmann AR et al. Cloning and characterisation of the rad9 DNA repair gene from Schizosaccharomyces pombe. Nucleic Acids Res 1991; 19:3525–3531.
34 Enoch T, Carr AM, Nurse P. Fission yeast genes involved in coupling mitosis to completion of DNA replication. Genes Dev 1992; 6:2035–2046.
35 al–Khodairy F, Carr AM. DNA repair mutants defining G2 checkpoint pathways in Schizosaccharomyces pombe. EMBO J 1992; 11:1343–1350.
36 Birkenbihl RP, Subramani S. Cloning and characterization of rad21 an essential gene of Schizosaccharomyces pombe involved in DNA double–strand–break repair. Nucleic Acids Res 1992; 20:6605–6611.
37 Jimenez G, Yucel J, Rowley R et al. The rad3+ gene of Schizosaccharomyces pombe is involved in multiple checkpoint functions and in DNA repair. Proc Natl Acad Sci USA 1992; 89:4952–4956.
38 Lieberman HB, Hopkins KM, Laverty M et al. Molecular cloning and analysis of Schizosaccharomyces pombe rad9, a gene involved in DNA repair and mutagenesis. Mol Gen Genet 1992; 232:367–376.
39 Rowley R, Subramani S, Young PG. Checkpoint controls in Schizosaccharomyces pombe: rad1. EMBO J 1992; 11:1335–1342.
40 al–Khodairy F, Fotou E, Sheldrick KS et al. Identification and characterization of new elements involved in checkpoint and feedback controls in fission yeast. Mol Biol Cell 1994; 5:147–160.
41 Saka Y, Fantes P, Yanagida M. Coupling of DNA replication and mitosis by fission yeast rad4/cut5. J Cell Sci Suppl 1994; 18:57–61.
42 Walworth N, Davey S, Beach D. Fission yeast chk1 protein kinase links the rad checkpoint pathway to cdc2. Nature 1993; 363:368–371.
43 Carr AM, Moudjou M, Bentley NJ et al. The chk1 pathway is required to prevent mitosis following cell–cycle arrest at 'start'. Curr Biol 1995; 5:1179–1190.
44 Walworth NC, Bernards R. rad–dependent response of the chk1–encoded protein kinase at the DNA damage checkpoint. Science 1996; 271:353–356.

45 Lydall D, Weinert T. From DNA damage to cell cycle arrest and suicide: a budding yeast perspective. Curr Opin Genet Dev 1996; 6:4–11.
46 Elledge SJ. Cell cycle checkpoints: preventing an identity crisis. Science 1996; 274:1664–1672.
47 Li FP, Fraumeni JF. Prospective study of a family cancer syndrome. JAMA 1982; 247:2692–2694.
48 Malkin D, Li FP, Strong LC et al. Germ line p53 mutations in a familial syndrome of breast cancer, sarcomas, and other neoplasms. Science 1990; 250:1233–1238.
49 Srivastava S, Zou ZQ, Pirollo K et al. Germ–line transmission of a mutated p53 gene in a cancer–prone family with Li–Fraumeni syndrome. Nature 1990; 348:747–749.
50 Maltzman W, Czyzyk L. UV irradiation stimulates levels of p53 cellular tumor antigen in nontransformed mouse cells. Mol Cell Biol 1984; 4:1689–1694.
51 Kastan MB, Onyekwere O, Sidransky D et al. Participation of p53 protein in the cellular response to DNA damage. Cancer Res 1991; 51:6304–6311.
52 Kastan MB, Zhan Q, el–Deiry WS et al. A mammalian cell cycle checkpoint pathway utilizing p53 and GADD45 is defective in ataxia–telangiectasia. Cell 1992; 71:587–597.
53 Kuerbitz SJ, Plunkett BS, Walsh WV et al. Wild–type p53 is a cell cycle checkpoint determinant following irradiation. Proc Natl Acad Sci USA 1992; 89:7491–7495.
54 Lane DP. Cancer. p53, guardian of the genome. Nature 1992; 358:15–16.
55 Lin D, Shields MT, Ullrich SJ et al. Growth arrest induced by wild–type p53 protein blocks cells prior to or near the restriction point in late G1 phase. Proc Natl Acad Sci USA 1992; 89:9210–9214.
56 Cohen MM, Levy HP. Chromosome instability syndromes. Adv Hum Genet 1989; 18:43–149, 365-371.
57 Savitsky K, Bar–Shira A, Gilad S et al. A single ataxia telangiectasia gene with a product similar to PI-3 kinase. Science 1995; 268:1749–1753.
58 Zampetti–Bosseler F, Scott D. Cell death, chromosome damage and mitotic delay in normal human, ataxia telangiectasia and retinoblastoma fibroblasts after x-irradiation. Int J Radiat Biol Relat Stud Phys Chem Med 1981; 39:547–558.
59 Nagasawa H, Latt SA, Lalande ME et al. Effects of X-irradiation on cell–cycle progression, induction of chromosomal aberrations and cell killing in ataxia telangiectasia (AT) fibroblasts. Mutat Res 1985; 148:71–82.
60 Rudolph NS, Latt SA. Flow cytometric analysis of X-ray sensitivity in ataxia telangiectasia. Mutat Res 1989; 211:31–41.
61 Lu X, Lane DP. Differential induction of transcriptionally active p53 following UV or ionizing radiation: defects in chromosome instability syndromes? Cell 1993; 75:765–778.
62 Lundgren K, Walworth N, Booher R et al. mik1 and wee1 cooperate in the inhibitory tyrosine phosphorylation of cdc2. Cell 1991; 64:1111–1122.
63 Enoch T, Nurse P. Mutation of fission yeast cell cycle control genes abolishes dependence of mitosis on DNA replication. Cell 1990; 60:665–673.
64 Saka Y, Fantes P, Sutani T et al. Fission yeast cut5 links nuclear chromatin and M phase regulator in the replication checkpoint control. EMBO J 1994; 13:5319–5329.
65 O'Connor PM, Ferris DK, Hoffmann I et al. Role of the cdc25C phosphatase in G2 arrest induced by nitrogen mustard. Proc Natl Acad Sci USA 1994; 91:9480–9484.
66 Smythe C, Newport JW. Coupling of mitosis to the completion of S phase in Xenopus occurs via modulation of the tyrosine kinase that phosphorylates $p34^{cdc2}$. Cell 1992; 68:787–797.
67 Kumagai A, Dunphy WG. Control of the Cdc2/cyclin B complex in Xenopus egg extracts arrested at a G2/M checkpoint with DNA synthesis inhibitors. Mol Biol Cell 1995; 6:199–213.
68 Rowley R, Hudson J, Young PG. The wee1 protein kinase is required for radiation–induced mitotic delay. Nature 1992; 356:353–355.
69 Rowley R. Radiation–induced mitotic delay: a genetic characterization in the fission yeast. Radiat Res 1992; 132:144–152.
70 Thiebaut F, Enns R, Howell SB. Cisplatin sensitivity correlates with its ability to cause cell cycle arrest via a wee1 kinase–dependent pathway in Schizosaccharomyces pombe. J Cell Physiol 1994; 159:506–514.
71 Barbet NC, Carr AM. Fission yeast wee1 protein kinase is not required for DNA damage–dependent mitotic arrest. Nature 1993; 364:824–827.

72 Herzinger T, Funk JO, Hillmer K et al. Ultraviolet B irradiation–induced G2 cell cycle arrest in human keratinocytes by inhibitory phosphorylation of the cdc2 cell cycle kinase. Oncogene 1995; 11:2151–2156.
73 el–Deiry WS, Tokino T, Velculescu VE et al. WAF1, a potential mediator of p53 tumor suppression. Cell 1993; 75:817–825.
74 Xiong Y, Zhang H, Beach D. D type cyclins associate with multiple protein kinases and the DNA replication and repair factor PCNA. Cell 1992; 71:505–514.
75 Zhang H, Xiong Y, Beach D. Proliferating cell nuclear antigen and p21 are components of multiple cell cycle kinase complexes. Mol Biol Cell 1993; 4:897–906.
76 Xiong Y, Zhang H, Beach D. Subunit rearrangement of the cyclin–dependent kinases is associated with cellular transformation. Genes Dev 1993; 7:1572–1583.
77 Xiong Y, Hannon GJ, Zhang H et al. p21 is a universal inhibitor of cyclin kinases. Nature 1993; 366:701–704.
78 Harper JW, Adami GR, Wei N et al. The p21 Cdk–interacting protein Cip1 is a potent inhibitor of G1 cyclin–dependent kinases. Cell 1993; 75:805–816.
79 Gu Y, Turck CW, Morgan DO. Inhibition of CDK2 activity in vivo by an associated 20K regulatory subunit. Nature 1993; 366:707–710.
80 Levine AJ. p53, the cellular gatekeeper for growth and division. Cell 1997; 88:323–331.
81 Noda A, Ning Y, Venable SF et al. Cloning of senescent cell–derived inhibitors of DNA synthesis using an expression screen. Exp Cell Res 1994; 211:90–98.
82 Dulic V, Kaufmann WK, Wilson SJ et al. p53–dependent inhibition of cyclin–dependent kinase activities in human fibroblasts during radiation–induced G1 arrest. Cell 1994; 76:1013-1023.
83 el–Deiry WS, Harper JW, O'Connor PM et al. WAF1/CIP1 is induced in p53–mediated G1 arrest and apoptosis. Cancer Res 1994; 54:1169–1174.
84 Waldman T, Kinzler KW, Vogelstein B. p21 is necessary for the p53–mediated G1 arrest in human cancer cells. Cancer Res 1995; 55:5187–5190.
85 Flores–Rozas H, Kelman Z, Dean FB et al. Cdk–interacting protein 1 directly binds with proliferating cell nuclear antigen and inhibits DNA replication catalyzed by the DNA polymerase delta holoenzyme. Proc Natl Acad Sci USA 1994; 91:8655–8659.
86 Li R, Waga S, Hannon GJ et al. Differential effects by the p21 CDK inhibitor on PCNA–dependent DNA replication and repair. Nature 1994; 371:534–537.
87 Waga S, Hannon GJ, Beach D et al. The p21 inhibitor of cyclin–dependent kinases controls DNA replication by interaction with PCNA. Nature 1994; 369:574–578.
88 Chen J, Jackson PK, Kirschner MW et al. Separate domains of p21 involved in the inhibition of Cdk kinase and PCNA. Nature 1995; 374:386–388.
89 Luo Y, Hurwitz J, Massague J. Cell–cycle inhibition by independent CDK and PCNA binding domains in $p21^{Cip1}$. Nature 1995; 375:159–161.
90 Nakanishi M, Robetorye RS, Pereira–Smith OM et al. The C-terminal region of $p21^{SDI1/WAF1/CIP1}$ is involved in proliferating cell nuclear antigen binding but does not appear to be required for growth inhibition. J Biol Chem 1995; 270:17060–17063.
91 Nakanishi M, Robetorye RS, Adami GR et al. Identification of the active region of the DNA synthesis inhibitory gene $p21^{Sdi1/CIP1/WAF1}$. EMBO J 1995; 14:555–563.
92 Podust VN, Podust LM, Goubin F et al. Mechanism of inhibition of proliferating cell nuclear antigen–dependent DNA synthesis by the cyclin–dependent kinase inhibitor p21. Biochemistry 1995; 34:8869–8875.
93 Chen IT, Akamatsu M, Smith ML et al. Characterization of $p21^{Cip1/Waf1}$ peptide domains required for cyclin E/Cdk2 and PCNA interaction. Oncogene 1996; 12:595–607.
94 Deng C, Zhang P, Harper JW et al. Mice lacking $p21^{CIP1/WAF1}$ undergo normal development, but are defective in G1 checkpoint control. Cell 1995; 82:675–684.
95 Brugarolas J, Chandrasekaran C, Gordon JI et al. Radiation–induced cell cycle arrest compromised by p21 deficiency. Nature 1995; 377:552–557.
96 Harper JW, Elledge SJ, Keyomarsi K et al. Inhibition of cyclin–dependent kinases by p21. Mol Biol Cell 1995; 6:387–400.
97 Kato JY, Matsuoka M, Polyak K et al. Cyclic AMP–induced G1 phase arrest mediated by an inhibitor ($p27^{Kip1}$) of cyclin–dependent kinase 4 activation. Cell 1994; 79:487–496.

98 Aprelikova O, Xiong Y, Liu ET. Both p16 and p21 families of cyclin–dependent kinase (CDK) inhibitors block the phosphorylation of cyclin–dependent kinases by the CDK–activating kinase. J Biol Chem 1995; 270:18195–18197.
99 Zhang H, Hannon GJ, Beach D. p21–containing cyclin kinases exist in both active and inactive states. Genes Dev 1994; 8:1750–1758.
100 Chen YQ, Cipriano SC, Arenkiel JM et al. Tumor suppression by $p21^{WAF1}$. Cancer Res 1995; 55:4536-4539.
101 Nakanishi M, Adami GR, Robetorye RS et al. Exit from G0 and entry into the cell cycle of cells expressing $p21^{Sdi1}$ antisense RNA. Proc Natl Acad Sci USA 1995; 92:4352–4356.
102 Guo K, Wang J, Andres V et al. MyoD–induced expression of p21 inhibits cyclin–dependent kinase activity upon myocyte terminal differentiation. Mol Cell Biol 1995; 15:3823–3829.
103 Halevy O, Novitch BG, Spicer DB et al. Correlation of terminal cell cycle arrest of skeletal muscle with induction of p21 by MyoD. Science 1995; 267:1018–1021.
104 Parker SB, Eichele G, Zhang P et al. p53–independent expression of $p21^{Cip1}$ in muscle and other terminally differentiating cells. Science 1995; 267:1024–1027.
105 Skapek SX, Rhee J, Spicer DB et al. Inhibition of myogenic differentiation in proliferating myoblasts by cyclin D1–dependent kinase. Science 1995; 267:1022–1024.
106 Elbendary A, Berchuck A, Davis P et al. Transforming growth factor beta 1 can induce CIP1/WAF1 expression independent of the p53 pathway in ovarian cancer cells. Cell Growth Differ 1994; 5:1301–1307.
107 Johnson M, Dimitrov D, Vojta PJ et al. Evidence for a p53–independent pathway for upregulation of SDI1/CIP1/WAF1/p21 RNA in human cells. Mol Carcinog 1994; 11:59-64.
108 Michieli P, Chedid M, Lin D et al. Induction of WAF1/CIP1 by a p53–independent pathway. Cancer Res 1994; 54:3391–3395.
109 Datto MB, Li Y, Panus JF et al. Transforming growth factor beta induces the cyclin–dependent kinase inhibitor p21 through a p53–independent mechanism. Proc Natl Acad Sci USA 1995; 92:5545–5549.
110 Li CY, Suardet L, Little JB. Potential role of WAF1/Cip1/p21 as a mediator of TGF–beta cytoinhibitory effect. J Biol Chem 1995; 270:4971–4974.
111 Chin YE, Kitagawa M, Su WC et al. Cell growth arrest and induction of cyclin–dependent kinase inhibitor $p21^{WAF1/CIP1}$ mediated by STAT1. Science 1996; 272:719–722.
112 Jakus J, Yeudall WA. Growth inhibitory concentrations of EGF induce p21 (WAF1/Cip1) and alter cell cycle control in squamous carcinoma cells. Oncogene 1996; 12:2369–2376.
113 Zeng YX, el Deiry WS. Regulation of $p21^{WAF1/CIP1}$ expression by p53–independent pathways. Oncogene 1996; 12:1557–1564.
114 Shim J, Lee H, Park J et al. A non–enzymatic p21 protein inhibitor of stress–activated protein kinases. Nature 1996; 381:804-807.
115 Koff A, Ohtsuki M, Polyak K et al. Negative regulation of G1 in mammalian cells: inhibition of cyclin E-dependent kinase by TGF–beta. Science 1993; 260:536-539.
116 Polyak K, Lee MH, Erdjument–Bromage H et al. Cloning of $p27^{Kip1}$, a cyclin–dependent kinase inhibitor and a potential mediator of extracellular antimitogenic signals. Cell 1994; 78:59–66.
117 Polyak K, Kato JY, Solomon MJ et al. $p27^{Kip1}$, a cyclin–Cdk inhibitor, links transforming growth factor–beta and contact inhibition to cell cycle arrest. Genes Dev 1994; 8:9–22.
118 Slingerland JM, Hengst L, Pan CH et al. A novel inhibitor of cyclin–Cdk activity detected in transforming growth factor beta–arrested epithelial cells. Mol Cell Biol 1994; 14:3683–3694.
119 Firpo EJ, Koff A, Solomon MJ et al. Inactivation of a Cdk2 inhibitor during interleukin 2–induced proliferation of human T lymphocytes. Mol Cell Biol 1994; 14:4889–4901.
120 Nourse J, Firpo E, Flanagan WM et al. Interleukin–2–mediated elimination of the $p27^{Kip1}$ cyclin–dependent kinase inhibitor prevented by rapamycin. Nature 1994; 372:570–573.
121 Toyoshima H, Hunter T. p27, a novel inhibitor of G1 cyclin–Cdk protein kinase activity, is related to p21. Cell 1994; 78:67–74.
122 Coats S, Flanagan WM, Nourse J et al. Requirement of $p27^{Kip1}$ for restriction point control of the fibroblast cell cycle. Science 1996; 272:877–880.
123 Kranenburg O, Scharnhorst V, Van der Eb AJ et al. Inhibition of cyclin–dependent kinase activity triggers neuronal differentiation of mouse neuroblastoma cells. J Cell Biol 1995; 131:227–234.

124 Liu M, Lee MH, Cohen M et al. Transcriptional activation of the Cdk inhibitor p21 by vitamin D3 leads to the induced differentiation of the myelomonocytic cell line U937. Genes Dev 1996; 10:142–153.
125 Chen J, Willingham T, Shuford M et al. Tumor suppression and inhibition of aneuploid cell accumulation in human brain tumor cells by ectopic overexpression of the cyclin–dependent kinase inhibitor $p27^{KIP1}$. J Clin Invest 1996; 97:1983–1988.
126 Nakayama K, Ishida N, Shirane M et al. Mice lacking $p27^{Kip1}$ display increased body size, multiple organ hyperplasia, retinal dysplasia, and pituitary tumors. Cell 1996; 85:707–720.
127 Kiyokawa H, Kineman RD, Manova–Todorova KO et al. Enhanced growth of mice lacking the cyclin–dependent kinase inhibitor function of $p27^{Kip1}$. Cell 1996; 85:721–732.
128 Fero ML, Rivkin M, Tasch M et al. A syndrome of multiorgan hyperplasia with features of gigantism, tumorigenesis, and female sterility in $p27^{Kip1}$–deficient mice. Cell 1996; 85:733–744.
129 Eblen ST, Fautsch MP, Anders RA et al. Conditional binding to and cell cycle–regulated inhibition of cyclin–dependent kinase complexes by $p27^{Kip1}$. Cell Growth Differ 1995; 6:915–925.
130 Russo AA, Jeffrey PD, Patten AK et al. Crystal structure of the $p27^{Kip1}$ cyclin–dependent–kinase inhibitor bound to the cyclin A–Cdk2 complex. Nature 1996; 382:325–331.
131 Pagano M, Tam SW, Theodoras AM et al. Role of the ubiquitin–proteasome pathway in regulating abundance of the cyclin–dependent kinase inhibitor p27. Science 1995; 269:682–685.
132 Schwob E, Bohm T, Mendenhall MD et al. The B–type cyclin kinase inhibitor $p40^{SIC1}$ controls the G1 to S transition in S. cerevisiae. Cell 1994; 79:233–244.
133 Matsuoka S, Edwards MC, Bai C et al. $p57^{KIP2}$, a structurally distinct member of the $p21^{CIP1}$ Cdk inhibitor family, is a candidate tumor suppressor gene. Genes Dev 1995; 9:650–662.
134 Lee MH, Reynisdottir I, Massague J. Cloning of $p57^{KIP2}$, a cyclin–dependent kinase inhibitor with unique domain structure and tissue distribution. Genes Dev 1995; 9:639–649.
135 Serrano M, Hannon GJ, Beach D. A new regulatory motif in cell–cycle control causing specific inhibition of cyclin D/CDK4. Nature 1993; 366:704–707.
136 Bork P. Hundreds of ankyrin–like repeats in functionally diverse proteins: mobile modules that cross phyla horizontally? Proteins 1993; 17:363–374.
137 Hannon GJ, Beach D. $p15^{INK4B}$ is a potential effector of TGF–beta–induced cell cycle arrest. Nature 1994; 371:257–261.
138 Bates S, Parry D, Bonetta L et al. Absence of cyclin D/cdk complexes in cells lacking functional retinoblastoma protein. Oncogene 1994; 9:1633–1640.
139 Parry D, Bates S, Mann DJ et al. Lack of cyclin D-Cdk complexes in Rb–negative cells correlates with high levels of $p16^{INK4/MTS1}$ tumour suppressor gene product. EMBO J 1995; 14:503–511.
140 Kamb A, Gruis NA, Weaver–Feldhaus J et al. A cell cycle regulator potentially involved in genesis of many tumor types. Science 1994; 264:436–440.
141 Guan KL, Jenkins CW, Li Y et al. Growth suppression by p18, a $p16^{INK4/MTS1}$– and $p14^{INK4B/MTS2}$–related CDK6 inhibitor, correlates with wild–type pRb function. Genes Dev 1994; 8:2939–2952.
142 Lukas J, Parry D, Aagaard L et al. Retinoblastoma–protein–dependent cell–cycle inhibition by the tumour suppressor p16. Nature 1995; 375:503–506.
143 Serrano M, Gomez–Lahoz E, DePinho RA et al. Inhibition of ras–induced proliferation and cellular transformation by $p16^{INK4}$. Science 1995; 267:249–252.
144 Arap W, Nishikawa R, Furnari FB et al. Replacement of the p16/CDKN2 gene suppresses human glioma cell growth. Cancer Res 1995; 55:1351–1354.
145 Enders GH, Koh J, Missero C et al. p16 inhibition of transformed and primary squamous epithelial cells. Oncogene 1996; 12:1239–1245.
146 Okamoto A, Demetrick DJ, Spillare EA et al. Mutations and altered expression of $p16^{INK4}$ in human cancer. Proc Natl Acad Sci USA 1994; 91:11045–11049.
147 Spillare EA, Okamoto A, Hagiwara K et al. Suppression of growth in vitro and tumorigenicity in vivo of human carcinoma cell lines by transfected $p16^{INK4}$. Mol Carcinog 1996; 16:53–60.
148 Hara E, Smith R, Parry D et al. Regulation of $p16^{CDKN2}$ expression and its implications for cell immortalization and senescence. Mol Cell Biol 1996; 16:859–867.
149 Reznikoff CA, Yeager TR, Belair CD et al. Elevated p16 at senescence and loss of p16 at immortalization in human papillomavirus 16 E6, but not E7, transformed human uroepithelial cells. Cancer Res 1996; 56:2886–2890.

150 Reynisdottir I, Polyak K, Iavarone A et al. Kip/Cip and Ink4 Cdk inhibitors cooperate to induce cell cycle arrest in response to TGF–beta. Genes Dev 1995; 9:1831–1845.
151 Stone S, Dayananth P, Jiang P et al. Genomic structure, expression and mutational analysis of the P15 (MTS2) gene. Oncogene 1995; 11:987–991.
152 Chan FK, Zhang J, Cheng L et al. Identification of human and mouse p19, a novel CDK4 and CDK6 inhibitor with homology to p16^{ink4}. Mol Cell Biol 1995; 15:2682–2688.
153 Hirai H, Roussel MF, Kato JY et al. Novel INK4 proteins, p19 and p18, are specific inhibitors of the cyclin D-dependent kinases CDK4 and CDK6. Mol Cell Biol 1995; 15:2672–2681.
154 Guan KL, Jenkins CW, Li Y et al. Isolation and characterization of p19^{INK4d}, a p16–related inhibitor specific to CDK6 and CDK4. Mol Biol Cell 1996; 7:57–70.
155 Okuda T, Hirai H, Valentine VA et al. Molecular cloning, expression pattern, and chromosomal localization of human CDKN2D/INK4d, an inhibitor of cyclin D-dependent kinases. Genomics 1995; 29:623–630.
156 Reed SI. The selection of S. cerevisiae mutants defective in the start event of cell division. Genetics 1980; 95:561–577.
157 Gerber MR, Farrell A, Deshaies RJ et al. Cdc37 is required for association of the protein kinase Cdc28 with G1 and mitotic cyclins. Proc Natl Acad Sci USA 1995; 92:4651–4655.
158 Stepanova L, Leng X, Parker SB et al. Mammalian p50^{Cdc37} is a protein kinase–targeting subunit of Hsp90 that binds and stabilizes Cdk4. Genes Dev 1996; 10:1491–1502.
159 Dai K, Kobayashi R, Beach D. Physical interaction of mammalian CDC37 with CDK4. J Biol Chem 1996; 271:22030–22034.

Chapter 6

Effectors of Cyclin–CDK Complexes

Up to this point in our consideration of cell cycle control, much attention has been devoted to the particulars of cyclin–CDK regulation without regard for the roles the complexes play in the cellular environment. CDKs are virtually unique among kinases in that there are four constraints on their activity: the necessity of the presence of the cyclin subunit, of the proper phosphorylation states of two distinct sites of the kinase, and of the absence of inhibitor molecules. That the CDKs are so strictly regulated indicates that the timing of their activity is crucial for the cell's survival. Indeed, as we have seen, cyclin–CDKs can be thought of as the gatekeepers of the cell cycle, controlling such crucial transitions as the onset of mitosis, or the passage of the cell through the restriction (R) point and consequent commitment to a round of DNA replication and cell division.

Not surprisingly, investigators have devoted a considerable amount of time trying to elucidate exactly how cyclin–CDK complexes regulate the major steps in the growth cycle of the cell. This has entailed a determined search for the substrates of the kinases. In spite of all efforts, the list of putative downstream targets is as of yet limited in length, and only a few are understood in detail. Among the latter are the retinoblastoma tumor suppressor protein and its close relatives, p107 and p130. As it turns out, these proteins are of such importance in cell cycle control that we would do well to consider them regulators of comparable significance to the cyclins, CDKs and CKIs.

The Retinoblastoma Susceptibility Gene

Retinoblastoma is a rare malignant tumor in the eye occurring almost exclusively in young children. In a small percentage of cases the victim has a familial history of predisposition to retinoblastoma, but in most cases the tumor arises spontaneously. It was first proposed in 1971 that retinoblastoma arises from two independent mutations (the "two–hit hypothesis").[1] In familial cases, the first mutation is present in the germline and the second mutation is somatic, whereas in spontaneous cases both mutations are somatic. The retinoblastoma susceptibility gene, designated *RB1*, was isolated and shown to be mutated in retinoblastoma tumors;[2,3] indeed, two independent mutations are necessary because both alleles of *RB1* must be impaired for the tumor to arise.

Inactivation of both copies of *RB1* appears to trigger neoplastic growth in a number of cell types; mutations of *RB1* occur not only in retinoblastomas but in a variety of other tumors, including osteosarcomas, mesenchymal tumors, small cell lung carcinomas, breast carcinomas, bladder carcinomas, synovial sarcomas, prostate carcinomas, and leukemias.[2,4-13] Evidently the gene plays an important role in constraining the growth of normal cells. In its absence, cells appear to lose their ability to control their proliferative response to environmental signals. Because of its growth–inhibiting effect and the necessity of eliminating the activity of both alleles in order for tumorigenesis to take place, *RB1* is a classic example of the tumor suppressor gene.

The observation that *RB1* is mutated or deleted in a wide array of human tumors implies that it exerts growth regulatory effects in a wide range of tissues. Mouse models in which one or both of the *RB1* alleles have been replaced with defective alleles reveal the importance of the gene not only in cell cycle control but in developmental processes as well. Mice heterozygous for the *RB1* mutation develop pituitary tumors although, interestingly and inexplicably, they do not develop retinoblastomas.[14-16] Knockout mice homozygous for defective *RB1* die in utero at about the fourteenth day of gestation from abnormalities in the blood, liver and central nervous system.[14-16] Because the embryonic lethal phenotype of the homozygous knockout mouse precludes further analysis of the effects of *RB1* loss of function in other cell types, chimeric mice, i.e., mice partially composed of *RB1*–nullizygous cells, have been engineered using embryonic stem cells lacking both wild–type *RB1* alleles.[17,18] Astonishingly, up to 60% of the animal is composed of *RB1*–deficient cells; further, these cells are present in all examined tissues including the blood and central nervous system, hinting that *RB1* can act in a non–cell–autonomous fashion. Nevertheless, aberrant cells or aberrant cell proliferation is observed in the adrenal medulla, cerebellar cortex, liver, lens and retina.[17,18] Functional *RB1* is apparently crucial in some tissues in certain developmental contexts but unimportant in others, for which reason the gene is of interest to cell cycle researchers and developmental biologists alike.

Because of the breadth of the research that has been devoted to *RB1* in the last decade, we shall presently limit ourselves to only those aspects of *RB1*

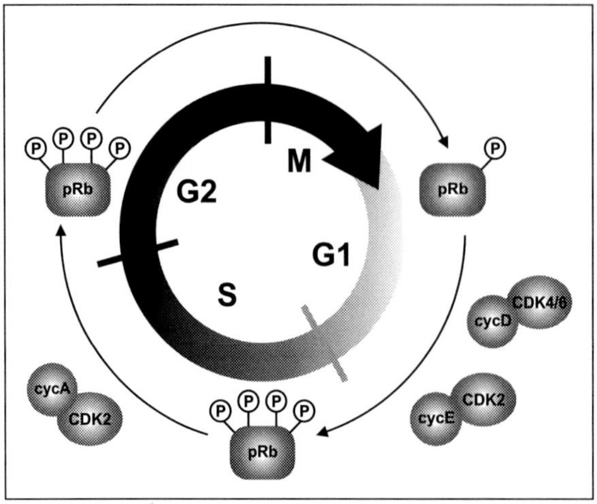

Fig. 6.1. Schematic illustration of the currently accepted model of pRb regulation during the cell cycle. In early G1 phase, pRb is found in a hypophosphorylated form. During mid–to–late G1 phase, cyclin D- and cyclin E-CDK complexes phosphorylate a number of sites (the four shown in the illustration are representative and do not reflect the actual number) on pRb, rendering the molecule into a hyperphosphorylated form. During S and G2 phases, the phosphorylation of the multiple residues in hyperphosphorylated pRb is maintained by cyclin A-CDK2. It is not until M phase, when cyclin A is degraded, that pRb becomes vulnerable to phosphatase action, which converts the molecule back into its hypophosphorylated form.

immediately relevant to cyclin–CDK function. Readers who are curious about this fascinating gene are directed to a number of review articles.[19-27]

The Retinoblastoma Tumor Suppressor Protein

RB1 encodes a 105 kDa nuclear phosphoprotein termed pRb.[28] It has been clearly demonstrated that while pRb levels show minimal fluctuation during the phases of the cell cycle, pRb is subject to regulation by phosphorylation in a cell cycle–dependent manner.[29-33] In early G1 phase and G0 phase, the protein is found in a hypophosphorylated form. As the cell traverses the R point and enters late G1 phase, the protein acquires phosphate groups on multiple serine and threonine residues; pRb becomes progressively more hyperphosphorylated through the extent of S phase, reaching a peak in G2 phase.[34-36] The phosphate groups have been removed by phosphatases by the time the cell reemerges into G1 phase after mitosis (Fig. 6.1).[37]

The available evidence suggests that cyclin–CDK complexes are the agents responsible for the phosphorylation of pRb as the cell cycle advances. CDK1, CDK2, CDK4 and CDK6 have all been observed to phosphorylate pRb in vitro.[34,38-47] Moreover, several studies in cell–based systems have demonstrated that certain combinations of cyclins and CDKs can phosphorylate pRb in vivo: cyclin A and CDK1, cyclin A and CDK2, cyclin E and CDK2, cyclin D (D1, D2, and D3) and CDK4, cyclin D and CDK6, but *not* cyclin B and CDK1.[40,45,48-51] The activities of these cyclin–CDK complexes peak at various times during G1, S and G2 phases, suggesting that each in turn participates in pRb phosphorylation according to the following scenario. Cyclin D-driven kinases act upon the hypophosphorylated form of pRb beginning in late G1; as these kinases are inactivated, cyclin E takes up the role of continuing the phosphorylation into S phase, at which point cyclin E is degraded and cyclin A assumes the responsibility of maintaining the hyperphosphorylated form of pRb through G2 phase. After cyclin A undergoes destruction in M phase, pRb becomes vulnerable to phosphatase action, restoring it to the hypophosphorylated form by early G1 phase (Fig. 6.1).

Plainly pRb is a crucial substrate if it is acted upon by so many cyclin–CDKs throughout such a long interval of the cell cycle. What purpose do the kinase complexes achieve by attaching phosphate groups to pRb? In exploring this question, we shall illuminate the role of pRb in cell cycle regulation.

pRb as a Regulator of Cellular Growth

The introduction of wild–type *RB1* alleles by viral infection, transfection, or microinjection into *RB1*–deficient cells derived from certain tumors induces growth arrest in these cells.[48,52-54] Clearly pRb is more than a tumor suppressor—it is a suppressor of proliferation in a general sense. It is significant that certain oncoproteins made by DNA tumor viruses when they infect cells—the E1A proteins of adenovirus; the E7 proteins of human papillomavirus; and the large T antigens of the polyoma family of viruses, particularly simian virus 40 (SV40)—physically associate with pRb; moreover, they form complexes almost exclusively with the hypophosphorylated, G1–specific form of pRb.[33,54-63] Binding of the viral proteins to pRb requires domains approximately consisting of amino acids 37-60 and 121-139 in E1A,[64,65] 18-37 in E7,[57] and 105-114 in SV40 large T antigen.[66,67] These domains are highly homologous to one another, and they all possess the common motif of Leu–X–Cys–X–Glu (X designates a variable amino acid). Complex formation with pRb seems to be critical to the ability of these oncoproteins to transform and immortalize the infected cells; when they are mutated in their pRb–binding domains they lose their oncogenic properties, and replacing one protein's domain with another's preserves these properties.[56,65,68-72]

Two conclusions may be drawn from these observations. First, pRb is of sufficient importance in growth regulation that viruses have evolved proteins that effectively remove it from the regulatory machinery of the cell; without these proteins, the viruses are unable to deregulate the cell cycle. Second, the oncoproteins specifically target hypophosphorylated pRb, suggesting that this form of pRb alone is active and possesses growth–suppressive properties. The hyperphosphorylated form, in contrast, is inactive (with regard to growth control) by virtue of its phosphorylation.

The idea that phosphorylation serves to inactivate pRb activity receives support from other studies. Through the use of cell fractionation techniques, it was observed that hypophosphorylated pRb binds to an unidentified "nuclear anchor," whereas the hyperphosphorylated form is unable to do so.[63,73] This binding appears to be prerequisite for pRb's growth–suppressing function, since mutant forms of pRb found in human tumors lack the ability to bind the anchor.[54,63]

What is the biochemical basis of pRb's growth inhibition, and how does phosphorylation obviate it? The identification of the domain through which pRb binds to the aforementioned oncoproteins provides an important clue. pRb interacts with these proteins by means of two nonadjacent regions approximately spanning amino acids 394–571 and 649–772; together the regions are known as the "pocket" domain.[67,74,75] An intact pocket domain is clearly essential to pRb's ability to inhibit proliferation, for mutations or deletions in the domain eliminate this ability.[54,76-78]

pRb has been observed to bind a number of molecules including a variety of stimulatory and repressive transcription factors,[79-90] kinases and phosphatases[91,92] and several proteins whose functions remain to be elucidated.[93-100] Most of these proteins are bound via the pocket domain, suggesting that the viral oncoproteins preempt pRb's growth–suppressive function by out competing its normal cellular partners. Tumor–derived mutations of pRb most often eliminate or alter this domain, and simultaneously prevent binding to these partners. Similarly, hyperphosphorylation of the pRb molecule may cause a conformational change such that access to the pocket domain is restricted and no binding can take place. Following this line of reasoning, one is led to the conclusion that pRb effects its anitproliferative program by sequestering proteins necessary for cell cycle progression (Fig. 6.2).

Among the cellular partners of pRb are the E2F family of transcription factors and, intriguingly, the D-type cyclins. These are the best understood of the pRb–binding proteins, and both are of critical importance in cell cycle regulation. As such we shall closely examine (or reexamine, in the case of cyclin D) each in turn. First, however, a few words must be said by way of introduction

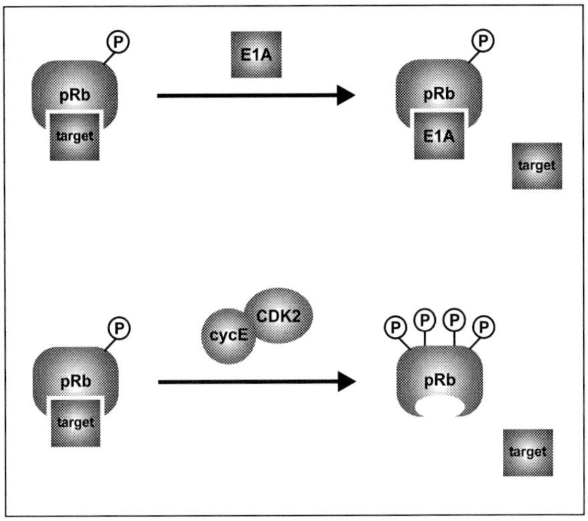

Fig. 6.2. Diagrams depicting two possible means by which the pRb molecule can be "inactivated." pRb contains a "pocket" domain through which it can bind to a number of transcription factors, kinases and phosphatases, and other proteins. When pRb exists in its hypophosphorylated form, the pocket domain assumes an "open" configuration that allows the binding of target proteins. pRb is believed to function as a suppressor of proliferation by virtue of its binding of downstream targets. Viral oncoproteins such as adenovirus E1A, SV40 large T antigen, and papillomavirus E7 can also bind pRb's pocket domain. Upon infection of a cell by adenovirus, for example, the E1A protein is produced and competitively binds pRb, causing the release of pRb's downstream targets and thereby "inactivating" pRb's inhibition of the cell cycle. Alternatively, cyclin–CDK complexes such as cyclin E–CDK2 can render pRb into a hyperphosphorylated form, causing the pocket domain to change conformation and assume a "closed" configuration. pRb is no longer able to bind target proteins and thereby becomes inactive.

about p107 and p130, a pair of proteins structurally and perhaps functionally related to pRb.

p107 and p130 were originally isolated as proteins that bind to viral oncoproteins such as adenovirus E1A and SV40 large T antigen in an identical fashion (i.e., via the same viral domains) as pRb.[65,101] Later work has established that both molecules bear structural kinship to pRb insofar as they possess homologous pocket domains (for which reason the three proteins as a group are known as the "pocket proteins"),[102-105] as well as functional kinship in that they display growth–suppressive properties distinct from but reminiscent of those of pRb.[106-108] Moreover, they are both nuclear phosphoproteins that are phosphorylated in a cell cycle–dependent manner, probably by cyclin–CDK complexes.[109-113] The numerous similarities to pRb suggest that p107 and p130 have related if not overlapping roles (although it is interesting to note that studies with knockout mice suggest that neither p107 nor p130 is individually essential for the appropriate regulation of the cell cycle).[114,115] As we shall see in the next section, this speculation is borne out by accumulating evidence that the two proteins are crucial co–participants along with pRb in the regulation of the E2F family of transcription factors.

E2F and DP Transcription Factors

E2F was first identified as a cellular transcription factor that binds to TTTCGCGC sequences in the adenovirus E2 promoter.[116] It had been well established that the E2 gene is induced by the E1A oncoprotein upon infection of cells by adenovirus; E2F provided the mechanistic explanation for this phenomenon. In uninfected cells, E2F resides in heteromeric complexes with other cellular proteins. After infection with adenovirus, E1A brings about the release of E2F from these complexes, and the liberated E2F effects the transcription of the E2 gene in concert with adenovirus–encoded E4.[117] An important clue concerning the nature of the complexes was provided by the discovery that pRb can bind to E2F in vivo and that E1A disrupts this binding.[79-82] Further, while the hypophosphorylated form of pRb is able to bind to E2F, the hyperphosphorylated form apparently lacks to ability to do so.[79,118,119]

From these observations one is led to suspect that pRb and E2F interact via the former's pocket domain. Studies have demonstrated that an intact pocket domain is indeed essential for E2F–binding to occur, but an extra stretch of amino acids C-terminal to the pocket is also necessary.[77,120] Interestingly, pRb is unable to enact growth suppression without both the pocket domain and the C-terminal domain, implying that interaction with E2F is the means through which pRb exerts its antiproliferative effects. Not surprisingly, complex formation with wild–type pRb silences E2F–mediated transcriptional activation; E2F can bind to pRb and DNA concurrently, indicating that pRb acts somewhat like a repressor domain attached to E2F.[120-125] Introducing viral oncoproteins such as E1A into the system negates this inhibition by pRb, as does the addition of cyclin–CDK complexes that phosphorylate pRb.[126,127] Plainly, E2F can switch from a transcriptional activator into a transcriptional repressor when it complexes with pRb and reverts to its original function when the complex is disrupted by a viral oncoprotein or pRb hyperphosphorylation (Fig. 6.3).

The importance of E2F in cell cycle control is signified on the one hand by the ability of overexpressed wild–type E2F to single–handedly drive quiescent cells into S phase, and on the other hand by the ability of dominant negative mutants of E2F to prevent cell cycle progression into S phase.[128-130] One might

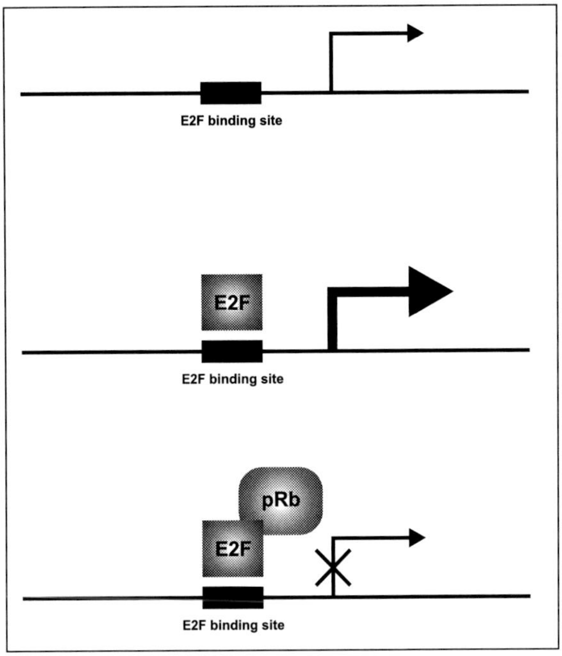

Fig. 6.3. Diagrams depicting the putative effects of pRb and E2F on the transcription of a gene with an E2F binding site in its promoter region. When neither E2F nor pRb is present, there is low–level basal transcription of the gene. When E2F alone is bound to the promoter, the E2F molecule acts as a transcriptional activator and substantially increases the transcription of the gene. When E2F binds the promoter and pRb simultaneously, the pRb–E2F complex acts as a transcriptional repressor and completely silences the transcription of the gene.

therefore suppose that pRb's primary role is to alter E2F activity in the cell. Consistent with this belief is a growing body of evidence suggesting that E2F induces a number of genes important in the passage of the cell through the G1/S phase transition as well as in the initiation of DNA synthesis. Among the putative targets of E2F transcription are cyclins D, E and A;[131-136] the enzymes dihydrofolate reductase, ribonucleotide reductase, thymidine kinase, thymidine synthase, and DNA polymerase alpha kinase;[132,137-144] and the products of the nuclear proto–oncogenes *MYC* and *BMYB*.[132,133,145-148] Moreover, mutant E2F that is unable to bind pRb can relieve cells of pRb–mediated growth suppression, lending support to the idea that E2F lies downstream of pRb.[149] Presumably pRb–E2F complexes repress E2F–responsive genes and block the cell cycle advance that would otherwise be facilitated by the transcription factor. E2F also appears to be able to induce transcription from its own promoter. This would provide an elegant mechanism for E2F autoregulation: prior to the G1/S phase transition, pocket protein–E2F complexes would repress the E2F gene and thereby establish a negative feedback loop; after the G1/S boundary (and pRb hyperphosphorylation), free E2F would stimulate a positive feedback loop and dramatically amplify its own expression.[150-153]

In actuality, E2F is almost never found bound solely to pRb in vivo; it is far more commonly observed to reside in ternary and quaternary complexes with additional proteins (Fig. 6.4). In recent years at least seven novel components found in various E2F heteromeric complexes have been cloned. To date investigators have characterized five human homologues of E2F.[118,154-161] A new family of E2F–binding proteins has also been identified, designated DP, of which there are at least two human homologues (DP–1 and DP–2).[162-165] A member of the E2F family is almost always found to be coupled to a member of the DP family in vivo; these heterodimers, when compared to the unbound E2F protein, have an enhanced ability to bind to DNA as well as to pRb and to regulate (whether positively or negatively) the transcription of target genes.[163,164,166,167] Investigators suspect that various heterodimers representing different combinations of E2Fs and DPs have distinct preferences for different groups of target genes. A proper evaluation of this conjecture awaits further study.

pRb's close cousins, p107 and p130, have also been implicated as regulatory components of E2F complexes (Fig. 6.4).[119,168-170] p107 in particular is known to act as a repressor domain in much the same fashion as pRb.[108] Each pocket protein prefers to bind to E2F during a distinct interval of the cell cycle: pRb in G1 and S phases, p107 in G1 and S phases and p130 in G0 and G1 phases.[119,170-173] In contrast to pRb, p107 and p130 have binding domains that allow them to interact with both cyclin A–CDK2 and cyclin E–CDK2 complexes. Yet they possess the ability to simultaneously bind E2F–DP dimers, allowing for the formation of complexes with five constituent proteins.[103,168-171,174,175] These complexes will form only after cyclin E-CDK2 and cyclin A–CDK2 have assembled, in late G1 phase and S phase respectively. Cyclin A–CDK2 has been observed in E2F complexes in the absence of a pocket proteins, and it apparently can abrogate E2F's DNA–binding ability by phosphorylating the DP subunit.[127,176-178] Because specific E2F molecules that preferentially bind p107 and p130 lack the ability to associate directly with cyclin A–CDK2, these pocket proteins are thought to act as mediators to bring E2F–DP and cyclin A–CDK2 together.[158-161,176-178] Regardless of the mechanistic details, we can speculate that an important role for cyclin A in S phase is the inactivation of E2F as a transcriptional regulator, a logical premise since all the available evidence suggests that the primary function of E2F is to promote the initiation of DNA synthesis at the G1/S phase transition. E2F activity, whether it takes the form of transcriptional activation or repression, is constrained to a defined interval spanning only the G1/S boundary. Cyclin E–

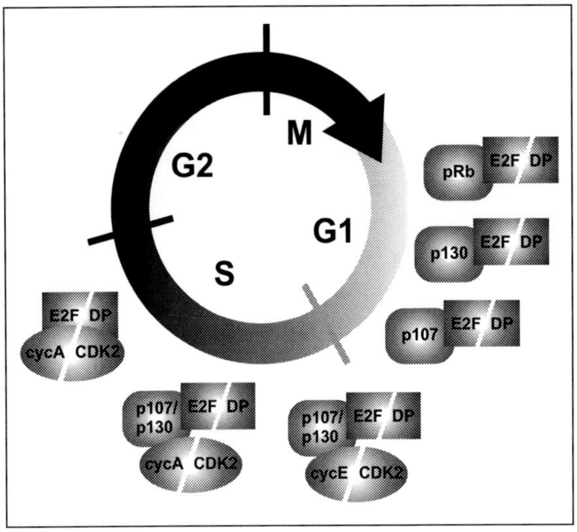

Fig. 6.4. Diagram depicting some of the numerous possible complexes containing a member of the E2F transcription factor family, as well as the intervals during the cell cycle when they are detected in the cell. Note that the complexes possess as many as five constituent proteins. The depictions of the complexes are not intended to accurately reflect the extent or lack of binding between various constituents of the complexes.

CDK2 has not been observed to act upon E2F–DP; what role cyclin E might play in E2F regulation is presently unclear.

Interestingly, the different E2F homologues preferentially bind to different pocket proteins. E2F–1, –2 and –3 bind exclusively to pRb,[157,179-181] and E2F–5 associates primarily with p130;[160] E2F–4 appears to interact with p130 during G1 phase but switches to pRb and p107 during S phase.[158,159,181,182] Clearly there is significant divergence among the homologues with regard to the means by which they are regulated, and we expect that there will be a corresponding divergence with regard to the targets of E2F regulation. Considerable investigation will be needed to flesh out the roles of the various E2F and DP proteins and their relationships with the cell cycle regulators.

We will conclude this section by emphasizing again the importance of E2F in the control of cell proliferation. It has recently become clear that E2F–1 (and potentially other members of the E2F family) possesses both of the seemingly contradictory roles of oncoprotein and tumor suppressor. Several studies have demonstrated that E2F has oncogenic properties that in the proper context can instigate neoplastic transformation.[183-186] Yet overexpressed E2F introduced into several cell lines can provoke programmed cell death, or apoptosis, apparently mediated through the p53 tumor suppressor protein (see next chapter).[129,178,187-190] Apoptosis is thought to serve as a protective mechanism that prevents a cell from being deregulated when it receives irreconcilable mitogenic and antiproliferative stimuli,[191] and E2F can be thought of as a cancer–inhibiting protein by virtue of its participation in this process. More telling is the recent characterization of E2F–1 knockout mice.[192,193] The mice are viable and healthy, indicating that E2F–1 is not essential for normal development or cell cycle progression, possibly because of redundant functions served by other E2F homologues. We might expect that the mice would display signs of underdevelopment because of E2F's role in proliferation, and indeed the mice experience testicular atrophy and exocrine gland dysplasia. Contrary to our expectation, however, as the mice age they develop tissues exhibiting hyperplastic growth as well as a range of tumors. These observations implicate E2F–1 as a tumor suppressor protein. How E2F–1 can be variously responsible for cell growth, the induction of apoptosis, and the suppression of oncogenesis is far from clear.[194] One conceivable explanation is that in the absence of E2F there are fewer pRb–E2F complexes to firmly repress S phase genes, and the basal transcription rate of these genes is sufficient to compromise the G1/S boundary checkpoint controls. Alternatively, without the contribution of E2F to cells' apoptotic defenses, perhaps not enough deregulated cells are eliminated to prevent tumor formation. A search for the answer to this puzzle will undoubtedly shed light on a number of aspects of cell cycle regulation.

The Interaction of pRb and D-Type Cyclins

The D-type cyclins are unique among the members of the cyclin family in that they are capable of stably binding to the pRb protein.[45,49,50] All three cyclin D isoforms contain the Leu–X–Cys–X–Glu motif shared by the viral oncoproteins that are known to bind pRb (adenovirus E1A, SV40 large T antigen, papillomavirus E7), and mutating this motif attenuates cyclin D-pRb binding.[50,195] Notably, cyclin D preferentially binds to the hypophosphorylated form of pRb, suggesting that it could act to inactivate pRb by out competing its downstream targets such as E2F.[50] At the same time, the D–type cyclins can partner with CDK4 and CDK6 and phosphorylate pRb, although it is unclear whether cyclin D–CDK complexes can render pRb into a fully hyperphosphorylated, inactive form in physiological conditions without the aid of other cyclin–CDK complexes.[45,49,195]

Is the principal cellular role of cyclin D to reverse pRb's suppression of proliferation by sequestering pRb away from its downstream targets, or to help obviate pRb function via phosphorylation (Fig. 6.5)? This question is still the subject of debate. The possibility exists that the binding of cyclin D to pRb allows the former to expeditiously bring CDK4 or CDK6 into proximity to pRb and

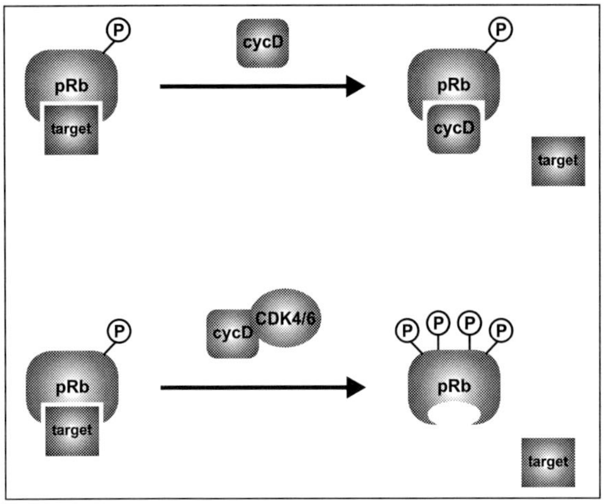

Fig. 6.5. Diagrams depicting two possible means by which the D-type cyclins might inactivate pRb. Each of the three isoforms of cyclin D can bind pRb via the latter's pocket domain in a manner similar to viral oncoproteins, suggesting that cyclin D may serve to release downstream targets of pRb through competitive binding. Cyclin D–CDK complexes can also phosphorylate (and presumably inactivate) pRb in a manner similar to cyclin E–CDK2 and cyclin A–CDK2, although it is not known whether cyclin D–CDKs can fully render pRb into its hyperphosphorylated form without the help of other cyclin–CDKs. Whether either one of these mechanisms or a combination of the two actually occurs in physiological conditions is unclear.

thereby facilitates the phosphorylation of pRb, although mutant cyclin D that is unable to bind pRb can still readily partner with CDKs and act upon pRb.[195] Regardless of how cyclin D inactivates pRb, there is strong evidence that pRb is the critical target of cyclin D action. Cyclin D1 function is dispensable for normal G1 phase progression in cells lacking pRb, while the introduction of pRb into these cells restores the need for cyclin D1.[196,197] Further, the growth–inhibitory CKI p16 (which specifically inhibits cyclin D–CDK complexes) requires the presence of cellular pRb in order to arrest the cell cycle.[198,199]

It cannot be ruled out that the function of cyclin D is not simply to bind pRb or to phosphorylate pRb, as suggested above, but involves other phenomena as well. It is possible that some hitherto unknown interaction between pRb and cyclin D can facilitate the progress of the cell cycle. Alternatively, cyclin D may work independently of pRb in some situations. In experiments with cultured cells, the addition of cyclin D1 in conjunction with oncogenes can cause cell transformation, and this ability seems to be dependent not on binding to pRb but rather on the activation of CDK4 and CDK6.[200] Since the phosphorylation state of pRb is unaltered in these experiments, one may conclude that cyclin D acts not through pRb inactivation but via some other substrate or substrates to promote cell proliferation. What these substrates might be is the subject of the following section.

Other Substrates of Cyclin–CDK Complexes

The reader should not be deceived by the almost exclusive attention given in this chapter to pRb into thinking that it is the only physiological cyclin–CDK substrate of significance. That we have devoted so much space to pRb, E2F and their relatives reflects the enormous wealth of research focused on these proteins during the past ten years. Indeed, they were the subjects of intense interest long before it was clear that they were intimately connected to cyclins and CDKs. There is little doubt that there exist a number of other important substrates of cyclin–CDK complexes; after all, the complexes control such critical checkpoints as the G1/S and G2/M phase transitions. The initiation of DNA replication and the onset of mitosis are events of extreme complexity that necessitate a coordination of cellular activities on a global rather than local scale. Since cyclin–CDKs provide the signals that set these processes into motions, it follows that they probably act upon a diverse array of target proteins in order for the signals to be propagated throughout the cell.

Finding these targets, however, is an enormously difficult undertaking. It is one thing to show that a cyclin–CDK complex can phosphorylate a potential substrate in vitro, and quite another thing to prove that the two interact in vivo in a physiologically important way. And even if the phosphorylation of a protein is linked to cyclin–CDK activity in vivo, one must demonstrate that there is a direct interaction between the proteins rather than one mediated by an intermediary kinase or a cascade of kinases. To further complicate the issue, because of the large respective memberships of the cyclin and CDK families, there are numerous possible cyclin–CDK combinations that must be tested on putative substrates since the various combinations are likely to possess different substrate specificities.

Over the years a host of potential substrates for cyclin A– and cyclin B–CDK1 have been identified, including chromatin components, nucleolar proteins, cytoskeletal proteins and microtubule–associated proteins. These candidates are not surprising since they are almost certainly involved in mitotic processes, but substantial work will be required to firmly establish them as bona fide substrates. The reader is directed to review articles for a comprehensive list of suspected targets of the mitotic cyclins.[201-203]

In contrast, the list of substrates for cyclin D– and cyclin E–CDK complexes remains short. The wealth of experimental evidence cited in this chapter suggests that pRb and probably p107 and p130 are the primary targets of the G1 phase cyclins. Recently a novel transcription factor, DMP1, has been identified as a likely substrate of cyclin D–CDK complexes; whether it has a role in cell cycle regulation or other genetic programs remains to be seen.[204] Cyclin E–CDK2, intriguingly, appears to phosphorylate cyclin E and thereby target the protein for ubiquitin–mediated proteolysis, establishing a negative feedback loop that elegantly explains the rapid degradation of cyclin E in early S phase.[205,206] A few other substrates have been proposed but not definitively proven.[203]

Disappointingly, the previous two paragraphs summarize the extent of our knowledge about CDK substrates at the present time. Reliable identification of additional CDK targets is a significant technical challenge and is likely to require innovative methodologies and years of hard work. And it will only be then that our understanding of the cell cycle regulators will become complete.

References

1 Knudson AG. Mutation and cancer: statistical study of retinoblastoma. Proc Natl Acad Sci USA 1971; 68:820–823.

2 Friend SH, Bernards R, Rogelj S et al. A human DNA segment with properties of the gene that predisposes to retinoblastoma and osteosarcoma. Nature 1986; 323:643–646.

3 Dunn JM, Phillips RA, Becker AJ et al. Identification of germline and somatic mutations affecting the retinoblastoma gene. Science 1988; 241:1797–1800.

4 Friend SH, Horowitz JM, Gerber MR et al. Deletions of a DNA sequence in retinoblastomas and mesenchymal tumors: organization of the sequence and its encoded protein. Proc Natl Acad Sci USA 1987; 84:9059–9063.

5 Harbour JW, Lai SL, Whang–Peng J et al. Abnormalities in structure and expression of the human retinoblastoma gene in SCLC. Science 1988; 241:353–357.

6 Lee EY, To H, Shew JY et al. Inactivation of the retinoblastoma susceptibility gene in human breast cancers. Science 1988; 241:218–221.

7 T'Ang A, Varley JM, Chakraborty S et al. Structural rearrangement of the retinoblastoma gene in human breast carcinoma. Science 1988; 242:263–266.

8 Varley JM, Armour J, Swallow JE et al. The retinoblastoma gene is frequently altered leading to loss of expression in primary breast tumours. Oncogene 1989; 4:725–729.

9 Horowitz JM, Yandell DW, Park SH et al. Point mutational inactivation of the retinoblastoma antioncogene. Science 1989; 243:937–940.

10 Shew JY, Ling N, Yang XM et al. Antibodies detecting abnormalities of the retinoblastoma susceptibility gene product (pp110^{RB}) in osteosarcomas and synovial sarcomas. Oncogene Res 1989; 4:205–214.

11 Horowitz JM, Park SH, Bogenmann E et al. Frequent inactivation of the retinoblastoma anti–oncogene is restricted to a subset of human tumor cells. Proc Natl Acad Sci USA 1990; 87:2775–2779.

12 Bookstein R, Rio P, Madreperla SA et al. Promoter deletion and loss of retinoblastoma gene expression in human prostate carcinoma. Proc Natl Acad Sci USA 1990; 87:7762–7766.

13 Furukawa Y, DeCaprio JA, Belvin M et al. Heterogeneous expression of the product of the retinoblastoma susceptibility gene in primary human leukemia cells. Oncogene 1991; 6:1343–1346.

14 Lee EY, Chang CY, Hu N et al. Mice deficient for Rb are nonviable and show defects in neurogenesis and haematopoiesis. Nature 1992; 359:288–294.
15 Jacks T, Fazeli A, Schmitt EM et al. Effects of an Rb mutation in the mouse. Nature 1992; 359:295–300.
16 Clarke AR, Maandag ER, van Roon M et al. Requirement for a functional Rb-1 gene in murine development. Nature 1992; 359:328–330.
17 Williams BO, Schmitt EM, Remington L et al. Extensive contribution of Rb–deficient cells to adult chimeric mice with limited histopathological consequences. EMBO J 1994; 13:4251–4259.
18 Maandag EC, van der Valk M, Vlaar M et al. Developmental rescue of an embryonic–lethal mutation in the retinoblastoma gene in chimeric mice. EMBO J 1994; 13:4260–4268.
19 Cobrinik D, Dowdy SF, Hinds PW et al. The retinoblastoma protein and the regulation of cell cycling. Trends Biochem Sci 1992; 17:312–315.
20 Hollingsworth RE, Hensey CE, Lee WH. Retinoblastoma protein and the cell cycle. Curr Opin Genet Dev 1993; 3:55–62.
21 Wang NP, To H, Lee WH et al. Tumor suppressor activity of RB and p53 genes in human breast carcinoma cells. Oncogene 1993; 8:279–288.
22 Zacksenhaus E, Bremner R, Jiang Z et al. Unraveling the function of the retinoblastoma gene. Adv Cancer Res 1993; 61:115–141.
23 Ewen ME. The cell cycle and the retinoblastoma protein family. Cancer Metastasis Rev 1994; 13:45–66.
24 Hinds PW, Weinberg RA. Tumor suppressor genes. Curr Opin Genet Dev 1994; 4:135–141.
25 Hinds PW. The retinoblastoma tumor suppressor protein. Curr Opin Genet Dev 1995; 5:79–83.
26 Weinberg RA. The retinoblastoma protein and cell cycle control. Cell 1995; 81:323–330.
27 Whyte P. The retinoblastoma protein and its relatives. Semin Cancer Biol 1995; 6:83–90.
28 Lee WH, Shew JY, Hong FD et al. The retinoblastoma susceptibility gene encodes a nuclear phosphoprotein associated with DNA binding activity. Nature 1987; 329:642–645.
29 DeCaprio JA, Ludlow JW, Lynch D et al. The product of the retinoblastoma susceptibility gene has properties of a cell cycle regulatory element. Cell 1989; 58:1085–1095.
30 Buchkovich K, Duffy LA, Harlow E. The retinoblastoma protein is phosphorylated during specific phases of the cell cycle. Cell 1989; 58:1097–1105.
31 Chen PL, Scully P, Shew JY et al. Phosphorylation of the retinoblastoma gene product is modulated during the cell cycle and cellular differentiation. Cell 1989; 58:1193–1198.
32 Mihara K, Cao XR, Yen A et al. Cell cycle–dependent regulation of phosphorylation of the human retinoblastoma gene product. Science 1989; 246:1300–1303.
33 Ludlow JW, Shon J, Pipas JM et al. The retinoblastoma susceptibility gene product undergoes cell cycle–dependent dephosphorylation and binding to and release from SV40 large T. Cell 1990; 60:387–396.
34 Lees JA, Buchkovich KJ, Marshak DR et al. The retinoblastoma protein is phosphorylated on multiple sites by human cdc2. EMBO J 1991; 10:4279–4290.
35 DeCaprio JA, Furukawa Y, Ajchenbaum F et al. The retinoblastoma–susceptibility gene product becomes phosphorylated in multiple stages during cell cycle entry and progression. Proc Natl Acad Sci USA 1992; 89:1795–1798.
36 Mittnacht S, Lees JA, Desai D et al. Distinct sub–populations of the retinoblastoma protein show a distinct pattern of phosphorylation. EMBO J 1994; 13:118–127.
37 Ludlow JW, Glendening CL, Livingston DM et al. Specific enzymatic dephosphorylation of the retinoblastoma protein. Mol Cell Biol 1993; 13:367–372.
38 Taya Y, Yasuda H, Kamijo M et al. In vitro phosphorylation of the tumor suppressor gene RB protein by mitosis–specific histone H1 kinase. Biochem Biophys Res Commun 1989; 164:580–586.
39 Lin BT, Gruenwald S, Morla AO et al. Retinoblastoma cancer suppressor gene product is a substrate of the cell cycle regulator cdc2 kinase. EMBO J 1991; 10:857–864.
40 Lin BT, Wang JY. Cell cycle regulation of retinoblastoma protein phosphorylation. Ciba Found Symp 1992; 170:227–241; discussion 241–243.
41 Hu QJ, Lees JA, Buchkovich KJ et al. The retinoblastoma protein physically associates with the human cdc2 kinase. Mol Cell Biol 1992; 12:971–980.

42 Williams RT, Carbonaro–Hall DA, Hall FL. Co–purification of p34^{cdc2}/p58$^{cyclin\ A}$ proline–directed protein kinase and the retinoblastoma tumor susceptibility gene product: interaction of an oncogenic serine/threonine protein kinase with a tumor–suppressor protein. Oncogene 1992; 7:423–432.
43 Akiyama T, Ohuchi T, Sumida S et al. Phosphorylation of the retinoblastoma protein by cdk2. Proc Natl Acad Sci USA 1992; 89:7900–7904.
44 Matsushime H, Ewen ME, Strom DK et al. Identification and properties of an atypical catalytic subunit (p34^{PSK-J3}/cdk4) for mammalian D type G1 cyclins. Cell 1992; 71:323–334.
45 Kato J, Matsushime H, Hiebert SW et al. Direct binding of cyclin D to the retinoblastoma gene product (pRb) and pRb phosphorylation by the cyclin D-dependent kinase CDK4. Genes Dev 1993; 7:331–342.
46 Meyerson M, Harlow E. Identification of G1 kinase activity for cdk6, a novel cyclin D partner. Mol Cell Biol 1994; 14:2077–2086.
47 Bates S, Bonetta L, MacAllan D et al. CDK6 (PLSTIRE) and CDK4 (PSK–J3) are a distinct subset of the cyclin–dependent kinases that associate with cyclin D1. Oncogene 1994; 9:71–79.
48 Hinds PW, Mittnacht S, Dulic V et al. Regulation of retinoblastoma protein functions by ectopic expression of human cyclins. Cell 1992; 70:993–1006.
49 Ewen ME, Sluss HK, Sherr CJ et al. Functional interactions of the retinoblastoma protein with mammalian D-type cyclins. Cell 1993; 73:487–497.
50 Dowdy SF, Hinds PW, Louie K et al. Physical interaction of the retinoblastoma protein with human D cyclins. Cell 1993; 73:499–511.
51 Horton LE, Qian Y, Templeton DJ. G1 cyclins control the retinoblastoma gene product growth regulation activity via upstream mechanisms. Cell Growth Differ 1995; 6:395–407.
52 Huang HJ, Yee JK, Shew JY et al. Suppression of the neoplastic phenotype by replacement of the RB gene in human cancer cells. Science 1988; 242:1563–1566.
53 Goodrich DW, Ping NW, Qian YW et al. The retinoblastoma gene product regulates progression through the G1 phase of the cell cycle. Cell 1991; 67:293–302.
54 Templeton DJ, Park SH, Lanier L et al. Nonfunctional mutants of the retinoblastoma protein are characterized by defects in phosphorylation, viral oncoprotein association, and nuclear tethering. Proc Natl Acad Sci USA 1991; 88:3033–3037.
55 Whyte P, Buchkovich KJ, Horowitz JM et al. Association between an oncogene and an anti–oncogene: the adenovirus E1A proteins bind to the retinoblastoma gene product. Nature 1988; 334:124–129.
56 DeCaprio JA, Ludlow JW, Figge J et al. SV40 large tumor antigen forms a specific complex with the product of the retinoblastoma susceptibility gene. Cell 1988; 54:275–283.
57 Dyson N, Howley PM, Munger K et al. The human papilloma virus–16 E7 oncoprotein is able to bind to the retinoblastoma gene product. Science 1989; 243:934–937.
58 Egan C, Bayley ST, Branton PE. Binding of the Rb1 protein to E1A products is required for adenovirus transformation. Oncogene 1989; 4:383–388.
59 Munger K, Werness BA, Dyson N et al. Complex formation of human papillomavirus E7 proteins with the retinoblastoma tumor suppressor gene product. EMBO J 1989; 8:4099–4105.
60 Ludlow JW, DeCaprio JA, Huang CM et al. SV40 large T antigen binds preferentially to an underphosphorylated member of the retinoblastoma susceptibility gene product family. Cell 1989; 56:57–65.
61 Dyson N, Bernards R, Friend SH et al. Large T antigens of many polyomaviruses are able to form complexes with the retinoblastoma protein. J Virol 1990; 64:1353–1356.
62 Imai Y, Matsushima Y, Sugimura T et al. Purification and characterization of human papillomavirus type 16 E7 protein with preferential binding capacity to the underphosphorylated form of retinoblastoma gene product. J Virol 1991; 65:4966–4972.
63 Mittnacht S, Hinds PW, Dowdy SF et al. Modulation of retinoblastoma protein activity during the cell cycle. Cold Spring Harb Symp Quant Biol 1991; 56:197–209.
64 Egan C, Jelsma TN, Howe JA et al. Mapping of cellular protein–binding sites on the products of early–region 1A of human adenovirus type 5. Mol Cell Biol 1988; 8:3955–3959.
65 Whyte P, Williamson NM, Harlow E. Cellular targets for transformation by the adenovirus E1A proteins. Cell 1989; 56:67–75.

66 Ewen ME, Ludlow JW, Marsilio E et al. An N–terminal transformation–governing sequence of SV40 large T antigen contributes to the binding of both $p110^{Rb}$ and a second cellular protein, p120. Cell 1989; 58:257–267.
67 Kaelin WG, Ewen ME, Livingston DM. Definition of the minimal simian virus 40 large T antigen– and adenovirus E1A–binding domain in the retinoblastoma gene product. Mol Cell Biol 1990; 10:3761–3769.
68 Moran E, Zerler B, Harrison TM et al. Identification of separate domains in the adenovirus E1A gene for immortalization activity and the activation of virus early genes. Mol Cell Biol 1986; 6:3470–3480.
69 Lillie JW, Loewenstein PM, Green MR et al. Functional domains of adenovirus type 5 E1a proteins. Cell 1987; 50:1091–1100.
70 Cherington V, Brown M, Paucha E et al. Separation of simian virus 40 large–T–antigen–transforming and origin–binding functions from the ability to block differentiation. Mol Cell Biol 1988; 8:1380–1384.
71 Smith DH, Ziff EB. The amino–terminal region of the adenovirus serotype 5 E1a protein performs two separate functions when expressed in primary baby rat kidney cells. Mol Cell Biol 1988; 8:3882–3890.
72 Moran E. A region of SV40 large T antigen can substitute for a transforming domain of the adenovirus E1A products. Nature 1988; 334:168–170.
73 Templeton DJ. Nuclear binding of purified retinoblastoma gene product is determined by cell cycle–regulated phosphorylation. Mol Cell Biol 1992; 12:435–443.
74 Hu QJ, Dyson N, Harlow E. The regions of the retinoblastoma protein needed for binding to adenovirus E1A or SV40 large T antigen are common sites for mutations. EMBO J 1990; 9:1147–1155.
75 Huang S, Wang NP, Tseng BY et al. Two distinct and frequently mutated regions of retinoblastoma protein are required for binding to SV40 T antigen. EMBO J 1990; 9:1815–1822.
76 Qian Y, Luckey C, Horton L et al. Biological function of the retinoblastoma protein requires distinct domains for hyperphosphorylation and transcription factor binding. Mol Cell Biol 1992; 12:5363–5372.
77 Qin XQ, Chittenden T, Livingston DM et al. Identification of a growth suppression domain within the retinoblastoma gene product. Genes Dev 1992; 6:953–964.
78 Hiebert SW. Regions of the retinoblastoma gene product required for its interaction with the E2F transcription factor are necessary for E2 promoter repression and pRb–mediated growth suppression. Mol Cell Biol 1993; 13:3384–3391.
79 Chellappan SP, Hiebert S, Mudryj M et al. The E2F transcription factor is a cellular target for the RB protein. Cell 1991; 65:1053–1061.
80 Bandara LR, La Thangue NB. Adenovirus E1a prevents the retinoblastoma gene product from complexing with a cellular transcription factor. Nature 1991; 351:494–497.
81 Chittenden T, Livingston DM, Kaelin WG. The T/E1A–binding domain of the retinoblastoma product can interact selectively with a sequence–specific DNA–binding protein. Cell 1991; 65:1073–1082.
82 Bagchi S, Weinmann R, Raychaudhuri P. The retinoblastoma protein copurifies with E2F-I, an E1A–regulated inhibitor of the transcription factor E2F. Cell 1991; 65:1063–1072.
83 Rustgi AK, Dyson N, Hill D et al. The c-myc oncoprotein forms a specific complex with the product of the retinoblastoma gene. Cold Spring Harb Symp Quant Biol 1991; 56:163–167.
84 Rustgi AK, Dyson N, Bernards R. Amino–terminal domains of c–myc and N–myc proteins mediate binding to the retinoblastoma gene product. Nature 1991; 352:541–544.
85 Kim SJ, Wagner S, Liu F et al. Retinoblastoma gene product activates expression of the human TGF–beta 2 gene through transcription factor ATF–2. Nature 1992; 358:331-334.
86 Gu W, Schneider JW, Condorelli G et al. Interaction of myogenic factors and the retinoblastoma protein mediates muscle cell commitment and differentiation. Cell 1993; 72:309–324.
87 Wang CY, Petryniak B, Thompson CB et al. Regulation of the Ets–related transcription factor Elf–1 by binding to the retinoblastoma protein. Science 1993; 260:1330–1335.

88 Hagemeier C, Bannister AJ, Cook A et al. The activation domain of transcription factor PU.1 binds the retinoblastoma (RB) protein and the transcription factor TFIID in vitro: RB shows sequence similarity to TFIID and TFIIB. Proc Natl Acad Sci USA 1993; 90:1580–1584.
89 Hagemeier C, Caswell R, Hayhurst G et al. Functional interaction between the HCMV IE2 transactivator and the retinoblastoma protein. EMBO J 1994; 13:2897–2903.
90 Iavarone A, Garg P, Lasorella A et al. The helix–loop–helix protein Id–2 enhances cell proliferation and binds to the retinoblastoma protein. Genes Dev 1994; 8:1270–1284.
91 Welch PJ, Wang JY. A C-terminal protein–binding domain in the retinoblastoma protein regulates nuclear c–Abl tyrosine kinase in the cell cycle. Cell 1993; 75:779–790.
92 Durfee T, Becherer K, Chen PL et al. The retinoblastoma protein associates with the protein phosphatase type 1 catalytic subunit. Genes Dev 1993; 7:555–569.
93 Kaelin WG, Pallas DC, DeCaprio JA et al. Identification of cellular proteins that can interact specifically with the T/E1A–binding region of the retinoblastoma gene product. Cell 1991; 64:521–532.
94 Huang S, Lee WH, Lee EY. A cellular protein that competes with SV40 T antigen for binding to the retinoblastoma gene product. Nature 1991; 350:160–162.
95 Defeo–Jones D, Huang PS, Jones RE et al. Cloning of cDNAs for cellular proteins that bind to the retinoblastoma gene product. Nature 1991; 352:251–254.
96 Fattaey AR, Helin K, Dembski MS et al. Characterization of the retinoblastoma binding proteins RBP1 and RBP2. Oncogene 1993; 8:3149–3156.
97 Qian YW, Wang YC, Hollingsworth RE et al. A retinoblastoma–binding protein related to a negative regulator of Ras in yeast. Nature 1993; 364:648–652.
98 Qian YW, Lee EY. Dual retinoblastoma–binding proteins with properties related to a negative regulator of ras in yeast. J Biol Chem 1995; 270:25507–25513.
99 Durfee T, Mancini MA, Jones D et al. The amino–terminal region of the retinoblastoma gene product binds a novel nuclear matrix protein that co–localizes to centers for RNA processing. J Cell Biol 1994; 127:609–622.
100 Chen PL, Ueng YC, Durfee T et al. Identification of a human homologue of yeast nuc2 which interacts with the retinoblastoma protein in a specific manner. Cell Growth Differ 1995; 652:199–210.
101 Dyson N, Buchkovich K, Whyte P et al. The cellular 107K protein that binds to adenovirus E1A also associates with the large T antigens of SV40 and JC virus. Cell 1989; 58:249–255.
102 Ewen ME, Xing YG, Lawrence JB et al. Molecular cloning, chromosomal mapping, and expression of the cDNA for p107, a retinoblastoma gene product–related protein. Cell 1991; 66:1155–1164.
103 Li Y, Graham C, Lacy S et al. The adenovirus E1A–associated 130–kD protein is encoded by a member of the retinoblastoma gene family and physically interacts with cyclins A and E. Genes Dev 1993; 7:2366–2377.
104 Hannon GJ, Demetrick D, Beach D. Isolation of the Rb–related p130 through its interaction with CDK2 and cyclins. Genes Dev 1993; 7:2378–2391.
105 Mayol X, Grana X, Baldi A et al. Cloning of a new member of the retinoblastoma gene family (pRb2) which binds to the E1A transforming domain. Oncogene 1993; 8:2561–2566.
106 Zhu L, van den Heuvel S, Helin K et al. Inhibition of cell proliferation by p107, a relative of the retinoblastoma protein. Genes Dev 1993, 7:1111–1125.
107 Claudio PP, Howard CM, Baldi A et al. p130/pRb2 has growth suppressive properties similar to yet distinctive from those of retinoblastoma family members pRb and p107. Cancer Res 1994; 54:5556–5560.
108 Starostik P, Chow KN, Dean DC. Transcriptional repression and growth suppression by the p107 pocket protein. Mol Cell Biol 1996; 16:3606–3614.
109 Beijersbergen RL, Carlee L, Kerkhoven RM et al. Regulation of the retinoblastoma protein–related p107 by G1 cyclin complexes. Genes Dev 1995; 9:1340–1353.
110 Baldi A, De Luca A, Claudio PP et al. The RB2/p130 gene product is a nuclear protein whose phosphorylation is cell cycle regulated. J Cell Biochem 1995; 59:402–408.
111 Mayol X, Garriga J, Grana X. Cell cycle–dependent phosphorylation of the retinoblastoma–related protein p130. Oncogene 1995; 11:801–808.

112 Claudio PP, De Luca A, Howard CM et al. Functional analysis of pRb2/p130 interaction with cyclins. Cancer Res 1996; 56:2003–2008.
113 Xiao ZX, Ginsberg D, Ewen M et al. Regulation of the retinoblastoma protein–related protein p107 by G1 cyclin–associated kinases. Proc Natl Acad Sci USA 1996; 93:4633–4637.
114 Lee MH, Williams BO, Mulligan G et al. Targeted disruption of p107: functional overlap between p107 and Rb. Genes Dev 1996; 10:1621–1632.
115 Cobrinik D, Lee MH, Hannon G et al. Shared role of the pRB–related p130 and p107 proteins in limb development. Genes Dev 1996; 10:1633–1644.
116 Kovesdi I, Reichel R, Nevins JR. Identification of a cellular transcription factor involved in E1A trans–activation. Cell 1986; 45:219–228.
117 Bagchi S, Raychaudhuri P, Nevins JR. Adenovirus E1A proteins can dissociate heteromeric complexes involving the E2F transcription factor: a novel mechanism for E1A trans–activation. Cell 1990; 62:659–669.
118 Kaelin WG, Krek W, Sellers WR et al. Expression cloning of a cDNA encoding a retinoblastoma–binding protein with E2F–like properties. Cell 1992; 70:351–364.
119 Shirodkar S, Ewen M, DeCaprio JA et al. The transcription factor E2F interacts with the retinoblastoma product and a p107–cyclin A complex in a cell cycle–regulated manner. Cell 1992; 68:157–166.
120 Hiebert SW, Chellappan SP, Horowitz JM et al. The interaction of RB with E2F coincides with an inhibition of the transcriptional activity of E2F. Genes Dev 1992; 6:177–185.
121 Hamel PA, Gill RM, Phillips RA et al. Transcriptional repression of the E2–containing promoters EIIaE, c–myc, and RB1 by the product of the RB1 gene. Mol Cell Biol 1992; 12:3431–3438.
122 Weintraub SJ, Prater CA, Dean DC. Retinoblastoma protein switches the E2F site from positive to negative element. Nature 1992; 358:259–261.
123 Helin K, Harlow E, Fattaey A. Inhibition of E2F–1 transactivation by direct binding of the retinoblastoma protein. Mol Cell Biol 1993; 13:6501–6508.
124 Sellers WR, Rodgers JW, Kaelin WG. A potent transrepression domain in the retinoblastoma protein induces a cell cycle arrest when bound to E2F sites. Proc Natl Acad Sci USA 1995; 92:11544–11548.
125 Weintraub SJ, Chow KN, Luo RX et al. Mechanism of active transcriptional repression by the retinoblastoma protein. Nature 1995; 375:812–815.
126 Zamanian M, La Thangue NB. Adenovirus E1a prevents the retinoblastoma gene product from repressing the activity of a cellular transcription factor. EMBO J 1992; 11:2603–2610.
127 Dynlacht BD, Flores O, Lees JA et al. Differential regulation of E2F transactivation by cyclin/cdk2 complexes. Genes Dev 1994; 8:1772–1786.
128 Johnson DG, Schwarz JK, Cress WD et al. Expression of transcription factor E2F1 induces quiescent cells to enter S phase. Nature 1993; 365:349–352.
129 Qin XQ, Livingston DM, Kaelin WG et al. Deregulated transcription factor E2F-1 expression leads to S-phase entry and p53–mediated apoptosis. Proc Natl Acad Sci USA 1994; 91:10918–10922.
130 Dobrowolski SF, Stacey DW, Harter ML et al. An E2F dominant negative mutant blocks E1A induced cell cycle progression. Oncogene 1994; 9:2605–2612.
131 Motokura T, Arnold A. PRAD1/cyclin D1 proto–oncogene: genomic organization, 5' DNA sequence, and sequence of a tumor–specific rearrangement breakpoint. Genes Chromosom Cancer 1993; 7:89–95.
132 Sala A, Nicolaides NC, Engelhard A et al. Correlation between E2F–1 requirement in the S phase and E2F–1 transactivation of cell cycle–related genes in human cells. Cancer Res 1994; 54:1402–1406.
133 DeGregori J, Kowalik T, Nevins JR. Cellular targets for activation by the E2F1 transcription factor include DNA synthesis– and G1/S–regulatory genes. Mol Cell Biol 1995; 15:4215–4224.
134 Duronio RJ, O'Farrell PH. Developmental control of the G1 to S transition in Drosophila: cyclin E is a limiting downstream target of E2F. Genes Dev 1995; 9:1456–1468.
135 Ohtani K, DeGregori J, Nevins JR. Regulation of the cyclin E gene by transcription factor E2F1. Proc Natl Acad Sci USA 1995; 92:12146–12150.
136 Schulze A, Zerfass K, Spitkovsky D et al. Cell cycle regulation of the cyclin A gene promoter is mediated by a variant E2F site. Proc Natl Acad Sci USA 1995; 92:11264–11268.

137 Blake MC, Azizkhan JC. Transcription factor E2F is required for efficient expression of the hamster dihydrofolate reductase gene in vitro and in vivo. Mol Cell Biol 1989; 9:4994–5002.
138 Mudryj M, Hiebert SW, Nevins JR. A role for the adenovirus inducible E2F transcription factor in a proliferation dependent signal transduction pathway. EMBO J 1990; 9:2179–2184.
139 Jolliff K, Li Y, Johnson LF. Multiple protein–DNA interactions in the TATAA–less mouse thymidylate synthase promoter. Nucleic Acids Res 1991; 19:2267–2274.
140 Pearson BE, Nasheuer HP, Wang TS. Human DNA polymerase alpha gene: sequences controlling expression in cycling and serum–stimulated cells. Mol Cell Biol 1991; 11:2081–2095.
141 Ogris E, Rotheneder H, Mudrak I et al. A binding site for transcription factor E2F is a target for trans activation of murine thymidine kinase by polyomavirus large T antigen and plays an important role in growth regulation of the gene. J Virol 1993; 67:1765–1771.
142 Slansky JE, Li Y, Kaelin WG et al. A protein synthesis–dependent increase in E2F1 mRNA correlates with growth regulation of the dihydrofolate reductase promoter. Mol Cell Biol 1993; 13:1610–1618.
143 Dou QP, Zhao S, Levin AH et al. G1/S–regulated E2F–containing protein complexes bind to the mouse thymidine kinase gene promoter. J Biol Chem 1994; 269:1306–1313.
144 Li Y, Slansky JE, Myers DJ et al. Cloning, chromosomal location, and characterization of mouse E2F1. Mol Cell Biol 1994; 14:1861–1869.
145 Hiebert SW, Lipp M, Nevins JR. E1A–dependent trans–activation of the human MYC promoter is mediated by the E2F factor. Proc Natl Acad Sci USA 1989; 86:3594–3598.
146 Huang MM, Hearing P. The adenovirus early region 4 open reading frame 6/7 protein regulates the DNA binding activity of the cellular transcription factor, E2F, through a direct complex. Genes Dev 1989; 3:1699–1710.
147 Thalmeier K, Synovzik H, Mertz R et al. Nuclear factor E2F mediates basic transcription and trans–activation by E1a of the human MYC promoter. Genes Dev 1989; 3:527–536.
148 Lam EW, Bennett JD, Watson RJ. Cell–cycle regulation of human B-myb transcription. Gene 1995; 160:277–281.
149 Qin XQ, Livingston DM, Ewen M et al. The transcription factor E2F-1 is a downstream target of RB action. Mol Cell Biol 1995; 15:742–755.
150 Hsiao KM, McMahon SL, Farnham PJ. Multiple DNA elements are required for the growth regulation of the mouse E2F1 promoter. Genes Dev 1994; 8:1526–1537.
151 Johnson DG, Ohtani K, Nevins JR. Autoregulatory control of E2F1 expression in response to positive and negative regulators of cell cycle progression. Genes Dev 1994; 8:1514–1525.
152 Johnson DG. Regulation of E2F-1 gene expression by p130 (Rb2) and D-type cyclin kinase activity. Oncogene 1995; 11:1685–1692.
153 Neuman E, Flemington EK, Sellers WR et al. Transcription of the E2F-1 gene is rendered cell cycle dependent by E2F DNA–binding sites within its promoter. Mol Cell Biol 1994; 14:6607–6615.
154 Helin K, Lees JA, Vidal M et al. A cDNA encoding a pRB–binding protein with properties of the transcription factor E2F. Cell 1992; 70:337–350.
155 Shan B, Zhu X, Chen PL et al. Molecular cloning of cellular genes encoding retinoblastoma–associated proteins: identification of a gene with properties of the transcription factor E2F. Mol Cell Biol 1992; 12:5620–5631.
156 Ivey–Hoyle M, Conroy R, Huber HE et al. Cloning and characterization of E2F–2, a novel protein with the biochemical properties of transcription factor E2F. Mol Cell Biol 1993; 13:7802–7812.
157 Lees JA, Saito M, Vidal M et al. The retinoblastoma protein binds to a family of E2F transcription factors. Mol Cell Biol 1993; 13:7813–7825.
158 Beijersbergen RL, Kerkhoven RM, Zhu L et al. E2F–4, a new member of the E2F gene family, has oncogenic activity and associates with p107 in vivo. Genes Dev 1994; 8:2680–2690.
159 Ginsberg D, Vairo G, Chittenden T et al. E2F-4, a new member of the E2F transcription factor family, interacts with p107. Genes Dev 1994; 8:2665–2679.
160 Hijmans EM, Voorhoeve PM, Beijersbergen RL et al. E2F-5, a new E2F family member that interacts with p130 in vivo. Mol Cell Biol 1995; 1555:3082–3089.

161 Sardet C, Vidal M, Cobrinik D et al. E2F–4 and E2F–5, two members of the E2F family, are expressed in the early phases of the cell cycle. Proc Natl Acad Sci USA 1995; 92:2403–2407.
162 Girling R, Partridge JF, Bandara LR et al. A new component of the transcription factor DRTF1/E2F. Nature 1993; 362:83–87.
163 Helin K, Wu CL, Fattaey AR et al. Heterodimerization of the transcription factors E2F–1 and DP–1 leads to cooperative trans–activation. Genes Dev 1993; 7:1850–1861.
164 Wu CL, Zukerberg LR, Ngwu C et al. In vivo association of E2F and DP family proteins. Mol Cell Biol 1995; 15:2536–2546.
165 Zhang Y, Chellappan SP. Cloning and characterization of human DP2, a novel dimerization partner of E2F. Oncogene 1995; 10:2085–2093.
166 Bandara LR, Buck VM, Zamanian M et al. Functional synergy between DP–1 and E2F–1 in the cell cycle–regulating transcription factor DRTF1/E2F. EMBO J 1993; 12:4317–4324.
167 Krek W, Livingston DM, Shirodkar S. Binding to DNA and the retinoblastoma gene product promoted by complex formation of different E2F family members. Science 1993; 262:1557–1560.
168 Cao L, Faha B, Dembski M et al. Independent binding of the retinoblastoma protein and p107 to the transcription factor E2F. Nature 1992; 355:176–179.
169 Devoto SH, Mudryj M, Pines J et al. A cyclin A-protein kinase complex possesses sequence–specific DNA binding activity: $p33^{cdk2}$ is a component of the E2F–cyclin A complex. Cell 1992; 68:167–176.
170 Cobrinik D, Whyte P, Peeper DS et al. Cell cycle–specific association of E2F with the p130 E1A–binding protein. Genes Dev 1993; 7:2392–2404.
171 Lees E, Faha B, Dulic V et al. Cyclin E/cdk2 and cyclin A/cdk2 kinases associate with p107 and E2F in a temporally distinct manner. Genes Dev 1992; 6:1874–1885.
172 Chittenden T, Livingston DM, DeCaprio JA. Cell cycle analysis of E2F in primary human T cells reveals novel E2F complexes and biochemically distinct forms of free E2F. Mol Cell Biol 1993; 13:3975–3983.
173 Schwarz JK, Devoto SH, Smith EJ et al. Interactions of the p107 and Rb proteins with E2F during the cell proliferation response. EMBO J 1993; 12:1013–1020.
174 Ewen ME, Faha B, Harlow E et al. Interaction of p107 with cyclin A independent of complex formation with viral oncoproteins. Science 1992; 255:85–87.
175 Faha B, Ewen ME, Tsai LH et al. Interaction between human cyclin A and adenovirus E1A–associated p107 protein. Science 1992; 255:87–90.
176 Xu M, Sheppard KA, Peng CY et al. Cyclin A/CDK2 binds directly to E2F–1 and inhibits the DNA–binding activity of E2F–1/DP–1 by phosphorylation. Mol Cell Biol 1994; 14:8420–8431.
177 Krek W, Ewen ME, Shirodkar S et al. Negative regulation of the growth–promoting transcription factor E2F–1 by a stably bound cyclin A–dependent protein kinase. Cell 1994; 78:161–172.
178 Krek W, Xu G, Livingston DM. Cyclin A–kinase regulation of E2F–1 DNA binding function underlies suppression of an S phase checkpoint. Cell 1995; 83:1149–1158.
179 Cress WD, Johnson DG, Nevins JR. A genetic analysis of the E2F1 gene distinguishes regulation by Rb, p107, and adenovirus E4. Mol Cell Biol 1993; 13:6314–6325.
180 Dyson N, Dembski M, Fattaey A et al. Analysis of p107–associated proteins: p107 associates with a form of E2F that differs from pRB–associated E2F-1. J Virol 1993; 67:7641–7647.
181 Moberg K, Starz MA, Lees JA. E2F–4 switches from p130 to p107 and pRB in response to cell cycle reentry. Mol Cell Biol 1996; 16:1436–1449.
182 Vairo G, Livingston DM, Ginsberg D. Functional interaction between E2F-4 and p130: evidence for distinct mechanisms underlying growth suppression by different retinoblastoma protein family members. Genes Dev 1995; 9:869–881.
183 Johnson DG, Cress WD, Jakoi L et al. Oncogenic capacity of the E2F1 gene. Proc Natl Acad Sci USA 1994; 91:12823–12827.
184 Singh P, Wong SH, Hong W. Overexpression of E2F-1 in rat embryo fibroblasts leads to neoplastic transformation. EMBO J 1994; 13:3329–3338.
185 Xu G, Livingston DM, Krek W. Multiple members of the E2F transcription factor family are the products of oncogenes. Proc Natl Acad Sci USA 1995; 92:1357–1361.
186 Yang XH, Sladek TL. Overexpression of the E2F–1 transcription factor gene mediates cell transformation. Gene Expr 1995; 4:195–204.

187 Shan B, Lee WH. Deregulated expression of E2F–1 induces S-phase entry and leads to apoptosis. Mol Cell Biol 1994; 14:8166–8173.
188 Wu X, Levine AJ. p53 and E2F–1 cooperate to mediate apoptosis. Proc Natl Acad Sci USA 1994; 91:3602–3606.
189 Kowalik TF, DeGregori J, Schwarz JK et al. E2F1 overexpression in quiescent fibroblasts leads to induction of cellular DNA synthesis and apoptosis. J Virol 1995; 69:2491–2500.
190 Shan B, Durfee T, Lee WH. Disruption of RB/E2F-1 interaction by single point mutations in E2F-1 enhances S-phase entry and apoptosis. Proc Natl Acad Sci USA 1996; 93:679–684.
191 Debbas M, White E. Wild–type p53 mediates apoptosis by E1A, which is inhibited by E1B. Genes Dev 1993; 7:546–554.
192 Yamasaki L, Jacks T, Bronson R et al. Tumor induction and tissue atrophy in mice lacking E2F–1. Cell 1996; 85:537–548.
193 Field SJ, Tsai FY, Kuo F et al. E2F–1 functions in mice to promote apoptosis and suppress proliferation. Cell 1996; 85:549–561.
194 Weinberg RA. E2F and cell proliferation: a world turned upside down. Cell 1996; 85:457–459.
195 Horton LE, Qian Y, Templeton DJ. G1 cyclins control the retinoblastoma gene product growth regulation activity via upstream mechanisms. Cell Growth Differ 1995; 6:395–407.
196 Lukas J, Muller H, Bartkova J et al. DNA tumor virus oncoproteins and retinoblastoma gene mutations share the ability to relieve the cell's requirement for cyclin D1 function in G1. J Cell Biol 1994; 125:625–638.
197 Lukas J, Bartkova J, Rohde M et al. Cyclin D1 is dispensable for G1 control in retinoblastoma gene–deficient cells independently of cdk4 activity. Mol Cell Biol 1995; 15:2600–2611.
198 Lukas J, Parry D, Aagaard L et al. Retinoblastoma–protein–dependent cell–cycle inhibition by the tumour suppressor p16. Nature 1995; 375:503–506.
199 Medema RH, Herrera RE, Lam F et al. Growth suppression by $p16^{ink4}$ requires functional retinoblastoma protein. Proc Natl Acad Sci USA 1995; 92:6289–6293.
200 Hinds PW, Dowdy SF, Eaton EN et al. Function of a human cyclin gene as an oncogene. Proc Natl Acad Sci USA 1994; 91:709–713.
201 Norbury C, Nurse P. Animal cell cycles and their control. Annu Rev Biochem 1992; 61:441–470.
202 Nigg EA. Targets of cyclin–dependent protein kinases. Curr Opin Cell Biol 1993; 5:187–193.
203 Nigg EA. Cyclin–dependent protein kinases: key regulators of the eukaryotic cell cycle. Bioessays 1995; 17:471–480.
204 Hirai H, Sherr CJ. Interaction of D-type cyclins with a novel myb–like transcription factor, DMP1. Mol Cell Biol 1996; 16:6457–6467.
205 Clurman BE, Sheaff RJ, Thress K et al. Turnover of cyclin E by the ubiquitin–proteasome pathway is regulated by cdk2 binding and cyclin phosphorylation. Genes Dev 1996; 10:1979–1990.
206 Won KA, Reed SI. Activation of cyclin E/CDK2 is coupled to site–specific autophosphorylation and ubiquitin–dependent degradation of cyclin E. EMBO J 1996; 15:4182–4193.

Chapter 7

Affectors of Cyclin–CDK Complexes

Now that we have discussed what is understood about the downstream targets of cyclin–CDK complexes, our cursory examination of the cell cycle machinery is concluded, for we have reached the boundary between what is known and what lies beyond the frontier of research. Nevertheless, it should be emphasized that although we have discussed cell cycle regulators as individual entities, they do not exist separate of their cellular milieu; their roles are defined in the context of their relationships with other components of the cell. Environmental signals originate outside the cell and are received at the plasma membrane. Receptors relay messages through signal transduction pathways into the interior of the cell. It is there that the signals are funneled into the cell cycle circuitry to be processed. Cyclins, CDKs, CKIs and their associated regulators compose the computer that decides whether the cell should proceed toward its proliferative destiny; but even the most sophisticated of computers is useless without streams of data to feed it.

An understanding of the pathways upstream of cyclin–CDK complexes is crucial if we are to do more than simply dissect the computer's circuits, if we are to discover ways to manipulate them to our advantage. For, as we shall see in the next chapter, investigators have come to suspect that the cell cycle machinery *must necessarily* be malfunctioning in order for cancer to arise. We can therefore employ one of two strategies in developing treatments for cancer: we can try to repair the computer, causing the cell to revert to a non–cancerous state; or we can try to shut down the computer, killing the cell. But at the present our therapies are limited to the use of exogenously delivered agents; we cannot directly intervene in a cell's internal affairs. It behooves us to seek an understanding of the signaling pathways linking the environment to the circuitry, for learning the system of "inputs" into the machinery may give us the means to "reprogram" it with the appropriate combination of pharmaceuticals and biological agents.

Unfortunately, this is the aspect of cell cycle regulation about which researchers know the least (although it is certain to become the subject of intense investigation in the near future). To date a few unambiguous connections between agonists and cell cycle regulators have been identified. For example, CD40 promotes B–cell proliferation by inducing the expression of CDK4 and CDK6,[1] and colony–stimulating factor-1 (CSF–1) exerts its mitogenic effects by upregulating cyclin D1 expression.[2-4] Nevertheless, the intermediary signaling proteins involved in these processes remain to be identified.

Although we are unable to offer a comprehensive description of the signaling pathways upstream of cyclin–CDKs, we will present three "case studies" that demonstrate the multitude of ways in which extracellular agents can affect the cell cycle regulators. The first examines DNA damage–induced cell cycle arrest mediated by the p53 tumor suppressor protein. The second describes the growth–suppressive effect of transforming growth factor–beta (TGF–β). The last considers the oncogenic activity of DNA tumor viruses. The important affectors of cyclin–CDKs described in these case studies will make clear both the versatility and the vulnerability of the cell cycle machinery.

The p53 Tumor Suppressor Protein

Like pRb, the p53 protein exhibits the properties of both a tumor suppressor and a growth suppressor. The p53 gene suffers mutations in a large number of human cancers, and wild–type *p53* can prevent the transformation of healthy cells by oncogenes and inhibit the growth of cells derived from tumors.[5-8] Moreover, mice that are nullizygous for wild–type *p53* are extremely susceptible to the development of malignant tumors (unlike pRb, p53 does not appear to play a crucial role in development and thus the knockout mice are viable).[9,10] As we mentioned in chapter 5, the p53 protein is believed to be the linchpin of a DNA–damage response pathway that causes the cell to arrest in late G1 phase, allowing the cell the opportunity for DNA repair before cell cycle progression resumes and DNA synthesis commences. Upon exposing a cell to DNA–damaging agents such as ionizing radiation, a rapid and substantial increase in the level of p53 is seen.[11,12] p53's pivotal role is underlined by the fact that *p53*–null cells cannot undergo cell cycle arrest following irradiation.[13]

p53 has also been observed to induce programmed cell death, or apoptosis, in a number of cell lines.[14-16] Thus p53 plays at least two important roles in preventing the cell's genetic material from accumulating undesirable oncogenic

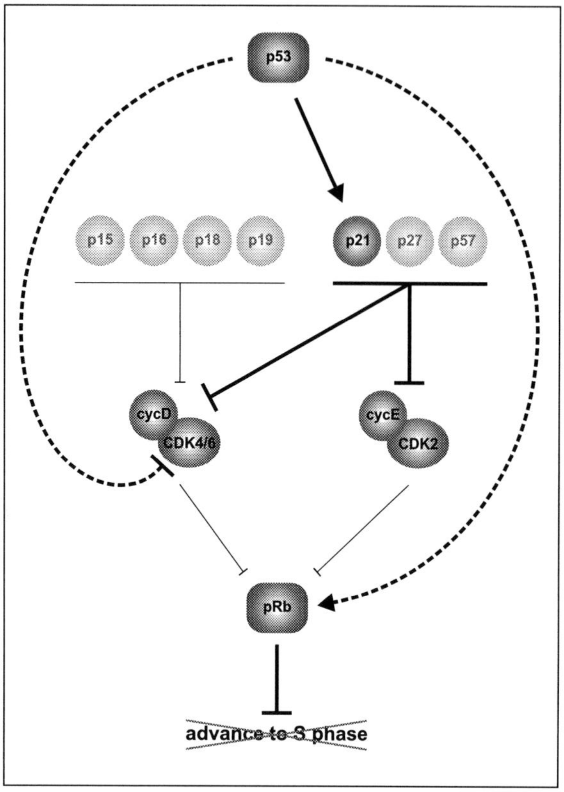

Fig. 7.1. Diagram depicting the cell cycle regulators through which the p53 tumor suppressor protein exerts its effects on cell proliferation. Arrows indicate an activation or upregulation of the target protein, and bars indicate an inhibition or downregulation of the target protein; thin lines indicate a *potential* interaction between the two proteins, and bold lines indicate an actualized interaction as a result of p53 activity; solid lines indicate that the interaction between the two proteins has been generally established by experimentation in a wide variety of systems, and dotted lines indicate that the interaction has been observed only in a limited number of systems. Overall, p53 prevents the cell from advancing into S phase.

mutations. If damage to the genome by harmful agents is moderate, p53 initiates growth arrest and DNA repair; if the damage is severe and irreparable, p53 mediates an apoptotic response that eliminates the threat of the deregulated cell to the organism. It is for these reasons that p53 has been dubbed the "guardian of the genome."[17]

Investigators are beginning to clarify how p53 erects a cell cycle block late in G1 phase (Fig. 7.1). An important clue to the inhibitory mechanism is provided by the observation that p53 can act as a transcription factor, promoting transcription from a defined DNA recognition site.[18-20] Implicit in this discovery is the conjecture that p53's transcriptional activity leads to the production of growth–suppressive proteins. In support of this idea, most p53 mutants derived from human tumors have lost their DNA–binding and transcription–activating powers.[18,21] On the flip side, p53 appears to be a potent inhibitor of the transcription of a wide array of cellular promoters by virtue of its ability to associate with the TATA–binding protein (TBP) component of the cell's transcriptional machinery.[22-25] Transcription cannot be initiated in the presence of p53, possibly because the machinery is prevented from binding to DNA.[24-27] It is conceivable that p53 inhibits the production of growth–promoting proteins.

As documented in the last chapter, one of the means by which the CKI p21 was discovered was through a screen for genes that are induced by overexpressed p53.[28] Simultaneously, p21 was identified as a protein that binds to and inhibits G1 phase cyclin–CDK complexes.[29-31] These findings suggested a compellingly simple model for p53–mediated growth suppression: p53 instigates the production of p21 via transcriptional activation of the CKI's gene, thereby obstructing cyclin D- and cyclin E-CDK activity, preventing the phosphorylation of pRb, and halting the cell at the restriction (R) point. This model has been borne out by studies showing that the irradiation of cells with wild–type p53, but not cells with mutant p53, causes increased p21 expression and leads to the accumulation of inactive cyclin E-CDK complexes.[32-34] Further, cells derived from *p21*–knockout mice have an impaired ability to respond to DNA–damaging agents with a late G1 phase arrest.[35,36] Additional supporting evidence that pRb lies directly downstream of p53 comes from the observation that the presence of viral oncoproteins that nullify pRb's growth–inhibitory abilities abrogates the cell cycle arrest engendered by p53.[37,38]

There are hints that p53 directly influences cell cycle regulators besides p21. One study reported that p53 can stimulate transcription from the human *RB1* promoter, suggesting that p53 may upregulate pRb in vivo.[39] This would stoichiometrically enhance the ability of pRb to sequester E2F transcription factor and would represent another means by which p53 could maintain the cell in G1 phase. Another study has observed that p53 may downregulate CDK4 by inhibiting its translation.[40] This would contribute to the impairment of cyclin D–CDK4 activity, keeping pRb in the growth–suppressive hypophosphorylated form. p53 is also known to induce the transcription of the cyclin G gene, but at the present time we can only speculate as to what the function of cyclin G in the cell and its relationship to p53 might be.[41,42] Additional targets of p53 may play a role in growth arrest by unknown mechanisms.

Parallel to the pathway by which p53 represses cell proliferation is an alternative pathway leading to apoptosis. The study of programmed cell death has emerged as a field in its own right and so lies beyond the scope of this book.

For our purposes it will suffice to say that apoptosis appears to be a wholly separate response that may not directly involve the cell cycle regulators we have discussed. Indeed, the present thinking is that p53–mediated G1 arrest and apoptosis represent mutually exclusive destinies between which a threatened cell must make a choice based on the severity of its condition. The interested reader is referred to recent reviews on the subject.[43-48]

Transforming Growth Factor–Beta (TGF–β)

TGF–β is unusual in that it has been shown to act in different contexts as either an inducer of cell proliferation or an inhibitor of cell growth.[49-51] While little is firmly known about how TGF–β gives rise to a proliferative response in certain cells, some progress has been made in identifying the means by which it generates antimitogenic signals in other cells. Not surprisingly, it has been linked to various cell cycle regulators.

Most telling is that TGF–β induces a late G1 phase cell cycle arrest when it is introduced to the appropriate cells at any time during early interphase.[52-54] This finding immediately suggests that it acts through G1 checkpoints, particularly the R point, which implicates cyclin D– and cyclin E–CDK complexes as the critical agents in the cell's response to TGF–β. This hypothesis is borne out by the observations that TGF–β prevents pRb phosphorylation in late G1 phase and that the growth–inhibitory effects of TGF–β are mitigated by the presence of viral oncoproteins that inactivate pRb's growth suppression.[52,53,55] Further, overexpression of E2F–1 transcription factor, the functional target of pRb, has been shown to relieve cells of TGF–β–mediated growth suppression.[56] pRb, as well as cyclin–CDKs (by virtue of their relationship to pRb), evidently lie downstream in the TGF–β signaling pathway.

There is evidence from studies of various cell lines that TGF–β impairs the expression of CDK2, CDK4, cyclin D and/or cyclin E.[40,57-60] Moreover, cyclin E–CDK2 complexes were found to be inactive in cells treated with TGF–β.[61] An explanation for this finding may lie in the observation that TGF–β enhances the activity of at least one of three CKIs—p15, p21 and p27—in various systems.[59,62-65] In mink cells, TGF–β induces the elevated expression of p15, which preferentially binds to CDK4 and CDK6; the increased p15 stoichiometrically displaces p27 from cyclin D–CDK4 and –CDK6 complexes; the released p27 disrupts cyclin E–CDK2 activity and prevents the cell from entering S phase.[65] Apparently, cyclin D–CDK complexes may be relatively "immune" to p27–mediated inhibition in some circumstances, and thus could serve as a "reservoir" for pre–formed p27. In this way, transcriptional induction of p15 by TGF–β signaling could rapidly eliminate CDK2, CDK4 and CDK6 activity and ensure a

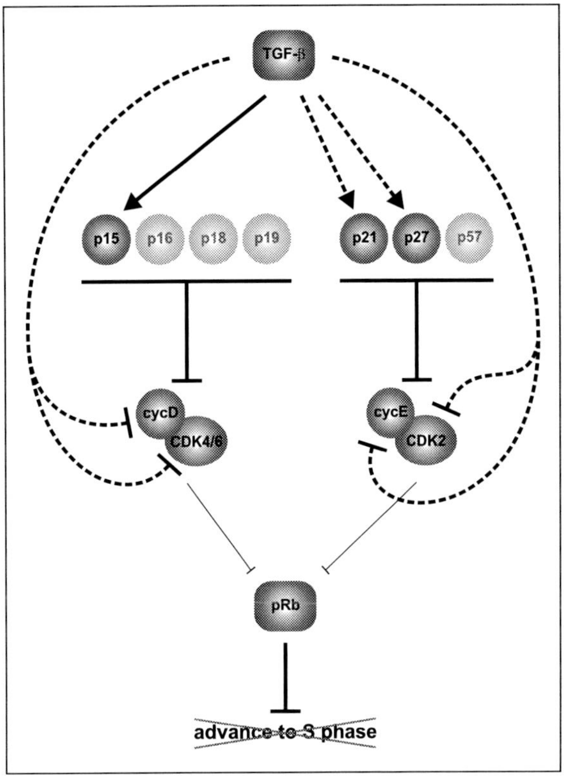

Fig. 7.2. Diagram depicting the cell cycle regulators through which transforming growth factor–beta (TGF–β) exerts its effects on cell proliferation. Arrows indicate an activation or upregulation of the target protein, and bars indicate an inhibition or downregulation of the target protein; thin lines indicate a *potential* interaction between the two proteins, and bold lines indicate an actualized interaction as a result of TGF–β exposure; solid lines indicate that the interaction between the two proteins has been generally established by experimentation in a wide variety of systems, and dotted lines indicate that the interaction has been observed only in a limited number of systems. Overall, TGF–β prevents the cell from advancing into S phase.

durable growth arrest. While this "inhibitor switching" mechanism provides an elegant explanation for the growth–suppressive effects of TGF–β, its necessity is cast into doubt by the finding that *p27*–nullizygous cells derived from knockout mice retain their susceptibility to TGF–β–mediated cell cycle arrest.

Regardless of the precise means, the multitudinous effects of TGF–β exposure on various cyclins, CDKs, and CKIs inhibit the formation of active cyclin–CDK complexes, leaving pRb in a hypophosphorylated state and indefinitely halting the cell in late G1 phase (Fig. 7.2).

DNA Tumor Viruses

Infection in animal models with any of three different tumor viruses—adenovirus, human papillomavirus and simian virus 40 (SV40)—has been observed to deregulate cell cycle progression. Indeed, high risk strains of papillomavirus are thought to play a contributory role in the genesis of virtually all cases of malignant human cervical cancer.[66-68] That these viruses should possess oncogenic properties seems counterproductive, for unrestrainedly proliferating cells are clearly not the most desirable hosts for infectious parasites. When we consider a virus' life cycle, however, we realize that its first priority is to exploit the host to synthesize many copies of the viral genome. A typical viral host might be a terminally differentiated cell locked in G0 phase, but the virus needs the cell to reside in S phase in order for it to access and commandeer the host's replicative machinery. The virus must therefore contrive to deregulate the checkpoints that hold the cell in quiescence and prevent its passage into S phase, particularly the R point and the G1/S phase transition. By throwing a wrench into the cell cycle machinery, however, the virus runs the risk of irrevocably unleashing proliferative forces that cause the cell to cycle out of control. If the virus engages in a lytic program this is not a concern, for the cell is killed before it is afforded the opportunity to reproduce. But in following a lysogenic program the virus allows the cell to proliferate, and in extraordinary circumstances the virus may inadvertently induce neoplastic growth.[69]

The tumor viruses push their replicative agenda upon host cells by means of virally encoded oncoproteins (Fig. 7.3). In the last chapter we mentioned three such proteins distinguished by their ability to form complexes with the pRb tumor suppressor protein and its relatives: adenovirus E1A, papillomavirus E7 and SV40 large T antigen. Each virus complements its pRb–binding oncoprotein with one that interacts with the p53 tumor suppressor protein: adenovirus E1B, papillomavirus E6, and SV40 large T antigen (which evidently serves the dual purpose).[70-72]

As we stated in the preceding chapter, the binding of viral oncoproteins to pRb (as well as p107 and p130) is essential to the oncogenic activities of the tumor viruses. Indeed, mutants of E7 which bind more tightly to pRb strengthen the transformation powers of papillomavirus.[73,74] This observation confirms our belief that the oncoproteins abrogate pRb function by displacing its normal cellular partners such as E2F from the pocket domain, achieving the same effect as hyperphosphorylation of pRb by cyclin–CDK complexes. The tumor viruses thereby bypass the regulatory pathways upstream of pRb and deceive host cells into inappropriately proceeding through pRb–mediated checkpoints.

p53 is the other significant obstacle to a virus–induced lapse into S phase, for as the "guardian of the genome" it can either cause the cell to arrest or in-

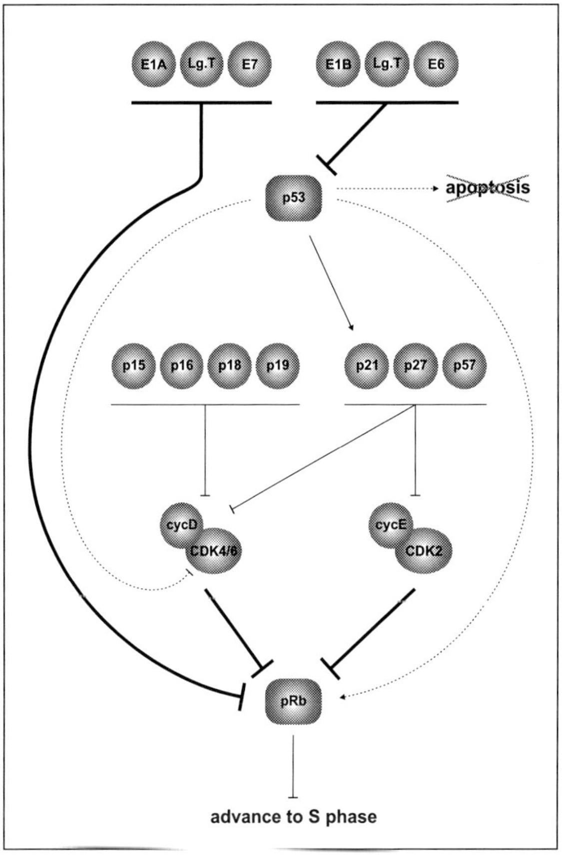

Fig. 7.3. Diagram depicting the cell cycle regulators through which the oncoproteins encoded by three tumor viruses—adenovirus (E1A and E1B proteins), simian virus 40 (large T antigen) and papillomavirus (E6 and E7 proteins)—exert their effects on cell proliferation. Arrows indicate an activation or upregulation of the target protein, and bars indicate an inhibition or downregulation of the target protein; thin lines indicate a *potential* interaction between the two proteins, and bold lines indicate an actualized interaction as a result of the presence of the viral oncoproteins; solid lines indicate that the interaction between the two proteins has been generally established by experimentation in a wide variety of systems, and dotted lines indicate that the interaction has been observed only in a limited number of systems. Overall, the viral oncoproteins enable the cell to advance into S phase and prevent apoptosis.

duce apoptosis if it detects some external threat. The p53–specific oncoproteins inhibit the tumor suppressor's functions by directly binding to the molecule,[75-80] although there is no single "pocket" of p53 as there is with pRb through which the interactions takes place. Accordingly, each oncoprotein counters p53's effects in a distinct fashion. E1B binds to the transcriptional activation domain, thereby acting as a transcriptional repressor of p53's target genes.[81,82] Large T antigen binds to the DNA–binding domain, preventing p53 from attaching to transcriptional sites and from fulfilling either its stimulatory or repressive activities as a transcription factor; the antigen also appears to stabilize p53 and increase its half–life.[83-85] E6 is similar to large T antigen in that it also binds to p53's DNA–binding domain, but the former contrasts with the latter in that it targets p53 for ubiquitin–mediated destruction and thus decreases the tumor suppressor protein's half–life.[78,86-88] Not only does E6's behavior have the effect of compromising the ability of p53 to prevent entry into S phase, it also prevents the host cell from mounting a p53–directed response to DNA damage.[89] The consequences are dire for the victimized cell, whose ability to repair damage and avoid the accumulation of mutations is impaired. The resultant genomic instability may explain why papillomavirus is a more potent carcinogenic agent than adenovirus or SV40.

Each of the three types of virus has the means to eliminate the growth–suppressive effects of the pRb and p53 proteins. What end does this achieve? p53 is crippled, no longer able to provoke an apoptotic response (which a virus must obviously avoid) or to induce cell cycle arrest by prompting the production of such factors as the CKI p21 and the repression of CDK4. With the absence of p21, a restraint on cyclin D– and cyclin E–CDK activity is removed. The cellular targets of CDK activity become more susceptible to phosphorylation, paving the way for the cell to bridge the R point and G1/S phase transition. pRb and its molecular cousins fail to sequester E2F and other transcription factors, allowing another portion of the cell's growth program to go forward. The cell is left with little in the way of resistance to being swept into S phase. Of potentially graver consequence to the greater organism is that the viruses knock out two crucial tumor suppressors, providing two "hits" out of the several mutations that are needed for a cell to emerge into a fully cancerous state (see chapter 8) and, in the case of papillomavirus, diminishing the ability of the cell to fend off more hits.

The Yin and Yang of Cell Cycle Regulation

Whereas p53 and TGF–β exploit cell cycle control proteins to inhibit cell cycle progression, the viral oncoproteins short–circuit the machinery to drive the cell cycle forward. These examples are an eminent demonstration of both the strengths and the weaknesses of the regulatory schematic with which cells are imprinted. There are a number of "positive" elements, such as cyclins and CDKs, interspersed with a number of "negative" elements, such as CKIs and pRb. The elements are set in opposition to each other, such that a tentative balance is achieved. The cell is enabled to respond appropriately and efficiently to both mitogenic and growth–inhibitory environmental signals; the former will reinforce the positive elements and weaken the negative elements, whereas the latter will strengthen the negative elements and repress the positive elements. Even if a cell receives conflicting signals, the stronger signal will prevail because the regulatory balance will be swayed in its favor.

The cell is thus endowed with exceptional responsiveness, but there is a price to pay. A few carefully placed strikes that eliminate key negative elements (e.g., the employment of oncoproteins by viruses) or that make key positive elements constitutively active (e.g., gain–of–function mutations) dramatically alter the balance in the cell. The positive elements can no longer be neutralized by the negative elements, and the result is continuous cell cycle progression. The cell becomes refractory to environmental signals. It has gone rogue, proliferating without regard to its neighbors or the organism as a whole. It is not hard to imagine cancer originating from a cell such as this.

Having examined the multiple facets of cyclin–CDK regulation as well as both some effectors and affectors of cyclin–CDK complexes, we have captured a snapshot of the field of cell cycle regulation more or less as it stands today. We are now in a position to frame the question of pressing interest to researchers and clinicians alike: what is the extent of the involvement of cell cycle regulators in the development of cancer? The answer, as we shall see in the next chapter, is that they are more closely involved that one might have suspected just a few years ago.

References

1 Ishida T, Kobayashi N, Tojo T et al. CD40 signaling–mediated induction of Bcl–x_L, Cdk4, and Cdk6. Implication of their cooperation in selective B cell growth. J Immunol 1995; 155:5527–5535.

2 Matsushime H, Roussel MF, Ashmun RA et al. Colony–stimulating factor 1 regulates novel cyclins during the G1 phase of the cell cycle. Cell 1991; 65:701–713.

3 Sherr CJ, Matsushime H, Roussel MF. Regulation of CYL/cyclin D genes by colony–stimulating factor 1. Ciba Found Symp 1992; 170:209–219; discussion 219–226.

4 Roussel MF, Theodoras AM, Pagano M et al. Rescue of defective mitogenic signaling by D–type cyclins. Proc Natl Acad Sci USA 1995; 92:6837–6841.

5 Finlay CA, Hinds PW, Levine AJ. The p53 proto–oncogene can act as a suppressor of transformation. Cell 1989; 57:1083–1093.

6 Eliyahu D, Michalovitz D, Eliyahu S et al. Wild–type p53 can inhibit oncogene–mediated focus formation. Proc Natl Acad Sci USA 1989; 86:8763–8767.

7 Baker SJ, Markowitz S, Fearon ER et al. Suppression of human colorectal carcinoma cell growth by wild–type p53. Science 1990; 249:912–915.

8 Johnson P, Gray D, Mowat M et al. Expression of wild–type p53 is not compatible with continued growth of p53–negative tumor cells. Mol Cell Biol 1991; 11:1–11.

9 Donehower LA, Harvey M, Slagle BL et al. Mice deficient for p53 are developmentally normal but susceptible to spontaneous tumours. Nature 1992; 356:215–221.

10 Kemp CJ, Donehower LA, Bradley A et al. Reduction of p53 gene dosage does not increase initiation or promotion but enhances malignant progression of chemically induced skin tumors. Cell 1993; 74:813–822.
11 Kastan MB, Onyekwere O, Sidransky D et al. Participation of p53 protein in the cellular response to DNA damage. Cancer Res 1991; 51:6304–6311.
12 Kuerbitz SJ, Plunkett BS, Walsh WV et al. Wild–type p53 is a cell cycle checkpoint determinant following irradiation. Proc Natl Acad Sci USA 1992; 89:7491–7495.
13 Kastan MB, Zhan Q, el–Deiry WS et al. A mammalian cell cycle checkpoint pathway utilizing p53 and GADD45 is defective in ataxia–telangiectasia. Cell 1992; 71:587–597.
14 Yonish–Rouach E, Resnitzky D, Lotem J et al. Wild–type p53 induces apoptosis of myeloid leukaemic cells that is inhibited by interleukin-6. Nature 1991; 352:345–347.
15 Shaw P, Bovey R, Tardy S et al. Induction of apoptosis by wild–type p53 in a human colon tumor–derived cell line. Proc Natl Acad Sci USA 1992; 89:4495–4499.
16 Ryan JJ, Danish R, Gottlieb CA et al. Cell cycle analysis of p53–induced cell death in murine erythroleukemia cells. Mol Cell Biol 1993; 13:711–719.
17 Lane DP. Cancer. p53, guardian of the genome. Nature 1992; 358:15–16.
18 el–Deiry WS, Kern SE, Pietenpol JA et al. Definition of a consensus binding site for p53. Nat Genet 1992; 1:45–49.
19 Zambetti GP, Bargonetti J, Walker K et al. Wild–type p53 mediates positive regulation of gene expression through a specific DNA sequence element. Genes Dev 1992; 6:1143–1152.
20 Funk WD, Pak DT, Karas RH et al. A transcriptionally active DNA–binding site for human p53 protein complexes. Mol Cell Biol 1992; 12:2866–2871.
21 Kern SE, Pietenpol JA, Thiagalingam S et al. Oncogenic forms of p53 inhibit p53–regulated gene expression. Science 1992; 256:827–830.
22 Truant R, Xiao H, Ingles CJ et al. Direct interaction between the transcriptional activation domain of human p53 and the TATA box–binding protein. J Biol Chem 1993; 286:2284–2287.
23 Martin DW, Munoz RM, Subler MA et al. p53 binds to the TATA–binding protein–TATA complex. J Biol Chem 1993; 268:13062–13067.
24 Liu X, Miller CW, Koeffler PH et al. The p53 activation domain binds the TATA box–binding polypeptide in Holo–TFIID, and a neighboring p53 domain inhibits transcription. Mol Cell Biol 1993; 13:3291–3300.
25 Seto E, Usheva A, Zambetti GP et al. Wild–type p53 binds to the TATA–binding protein and represses transcription. Proc Natl Acad Sci USA 1992; 89:12028–12032.
26 Ragimov N, Krauskopf A, Navot N et al. Wild–type but not mutant p53 can repress transcription initiation in vitro by interfering with the binding of basal transcription factors to the TATA motif. Oncogene 1993; 8:1183–1193.
27 Mack DH, Vartikar J, Pipas JM et al. Specific repression of TATA–mediated but not initiator–mediated transcription by wild–type p53. Nature 1993; 363:281–283.
28 el–Deiry WS, Tokino T, Velculescu VE et al. WAF1, a potential mediator of p53 tumor suppression. Cell 1993; 75:817–825.
29 Gu Y, Turck CW, Morgan DO. Inhibition of CDK2 activity in vivo by an associated 20K regulatory subunit. Nature 1993; 366:707–710.
30 Harper JW, Adami GR, Wei N et al. The p21 Cdk–interacting protein Cip1 is a potent inhibitor of G1 cyclin–dependent kinases. Cell 1993; 75:805–816.
31 Xiong Y, Hannon GJ, Zhang H et al. p21 is a universal inhibitor of cyclin kinases. Nature 1993; 366:701–704.
32 Dulic V, Kaufmann WK, Wilson SJ et al. p53–dependent inhibition of cyclin–dependent kinase activities in human fibroblasts during radiation–induced G1 arrest. Cell 1994; 76:1013–1023.
33 el–Deiry WS, Harper JW, O'Connor PM et al. WAF1/CIP1 is induced in p53–mediated G1 arrest and apoptosis. Cancer Res 1994; 54:1169–1174.
34 Liu M, Pelling JC. UV–B/A irradiation of mouse keratinocytes results in p53–mediated WAF1/CIP1 expression. Oncogene 1995; 10:1955–1960.
35 Deng C, Zhang P, Harper JW et al. Mice lacking $p21^{CIP1/WAF1}$ undergo normal development, but are defective in G1 checkpoint control. Cell 1995; 82:675–684.

36 Brugarolas J, Chandrasekaran C, Gordon JI et al. Radiation–induced cell cycle arrest compromised by p21 deficiency. Nature 1995; 377:552–557.
37 Demers GW, Foster SA, Halbert CL et al. Growth arrest by induction of p53 in DNA damaged keratinocytes is bypassed by human papillomavirus 16 E7. Proc Natl Acad Sci USA 1994; 91:4382–4386.
38 Slebos RJ, Lee MH, Plunkett BS et al. p53–dependent G1 arrest involves pRB–related proteins and is disrupted by the human papillomavirus 16 E7 oncoprotein. Proc Natl Acad Sci USA 1994; 91:5320–5324.
39 Osifchin NE, Jiang D, Ohtani–Fujita N et al. Identification of a p53 binding site in the human retinoblastoma susceptibility gene promoter. J Biol Chem 1994; 269:6383–6389.
40 Ewen ME, Oliver CJ, Sluss HK et al. p53–dependent repression of CDK4 translation in TGF–beta–induced G1 cell–cycle arrest. Genes Dev 1995; 9:204–217.
41 Okamoto K, Beach D. Cyclin G is a transcriptional target of the p53 tumor suppressor protein. EMBO J 1994; 13:4816–4822.
42 Zauberman A, Lupo A, Oren M. Identification of p53 target genes through immune selection of genomic DNA: the cyclin G gene contains two distinct p53 binding sites. Oncogene 1995; 10:2361–2366.
43 Evan GI, Brown L, Whyte M et al. Apoptosis and the cell cycle. Curr Opin Cell Biol 1995; 7:825–834.
44 Haffner R, Oren M. Biochemical properties and biological effects of p53. Curr Opin Genet Dev 1995; 5:84–90.
45 Kastan MB, Canman CE, Leonard CJ. P53, cell cycle control and apoptosis: implications for cancer. Cancer Metastasis Rev 1995; 14:3–15.
46 Lee JM, Bernstein A. Apoptosis, cancer and the p53 tumour suppressor gene. Cancer Metastasis Rev 1995; 14:149–161.
47 Liebermann DA, Hoffman B, Steinman RA. Molecular controls of growth arrest and apoptosis: p53–dependent and independent pathways. Oncogene 1995; 11:199–210.
48 Meikrantz W, Schlegel R. Apoptosis and the cell cycle. J Cell Biochem 1995; 58:160–174.
49 Moses HL, Yang EY, Pietenpol JA. TGF–beta stimulation and inhibition of cell proliferation: new mechanistic insights. Cell 1990; 63:245–247.
50 Roberts AB, Sporn MB. The transforming growth factor betas. In: Sporn MB, Roberts AB, eds. Peptide Growth Factors and their Receptors. Heidelberg: Springer–Verlag, 1990:419–472.
51 Massague J. The transforming growth factor–beta family. Annu Rev Cell Biol 1990; 6:597–641.
52 Pietenpol JA, Stein RW, Moran E et al. TGF–beta 1 inhibition of c-myc transcription and growth in keratinocytes is abrogated by viral transforming proteins with pRB binding domains. Cell 1990; 61:777–785.
53 Laiho M, DeCaprio JA, Ludlow JW et al. Growth inhibition by TGF–beta linked to suppression of retinoblastoma protein phosphorylation. Cell 1990; 62:175–185.
54 Howe PH, Draetta G, Leof EB. Transforming growth factor beta 1 inhibition of $p34^{cdc2}$ phosphorylation and histone H1 kinase activity is associated with G1/S–phase growth arrest. Mol Cell Biol 1991; 11:1185–1194.
55 Landesman Y, Pagano M, Draetta G et al. Modifications of cell cycle controlling nuclear proteins by transforming growth factor beta in the HaCaT keratinocyte cell line. Oncogene 1992; 7:1661–1665.
56 Schwarz JK, Bassing CH, Kovesdi I et al. Expression of the E2F1 transcription factor overcomes type beta transforming growth factor–mediated growth suppression. Proc Natl Acad Sci USA 1995; 92:483–487.
57 Ewen ME, Sluss HK, Whitehouse LL et al. TGF beta inhibition of Cdk4 synthesis is linked to cell cycle arrest. Cell 1993; 74:1009–1020.
58 Geng Y, Weinberg RA. Transforming growth factor beta effects on expression of G1 cyclins and cyclin–dependent protein kinases. Proc Natl Acad Sci USA 1993; 90:10315–10319.
59 Slingerland JM, Hengst L, Pan CH et al. A novel inhibitor of cyclin–Cdk activity detected in transforming growth factor beta–arrested epithelial cells. Mol Cell Biol 1994; 14:3683–3694.

60 Satterwhite DJ, Aakre ME, Gorska AE et al. Inhibition of cell growth by TGF beta 1 is associated with inhibition of B–myb and cyclin A in both BALB/MK and Mv1Lu cells. Cell Growth Differ 1994; 5:789–799.
61 Koff A, Ohtsuki M, Polyak K et al. Negative regulation of G1 in mammalian cells: inhibition of cyclin E–dependent kinase by TGF–beta. Science 1993; 260:536–539.
62 Polyak K, Kato JY, Solomon MJ et al. $p27^{Kip1}$, a cyclin–Cdk inhibitor, links transforming growth factor–beta and contact inhibition to cell cycle arrest. Genes Dev 1994; 8:9–22.
63 Hannon GJ, Beach D. $p15^{INK4B}$ is a potential effector of TGF–beta–induced cell cycle arrest. Nature 1994; 371:257–261.
64 Datto MB, Li Y, Panus JF et al. Transforming growth factor beta induces the cyclin–dependent kinase inhibitor p21 through a p53–independent mechanism. Proc Natl Acad Sci USA 1995; 92:5545–5549.
65 Reynisdottir I, Polyak K, Iavarone A et al. Kip/Cip and Ink4 Cdk inhibitors cooperate to induce cell cycle arrest in response to TGF–beta. Genes Dev 1995; 9:1831–1845.
66 Parkin DM, Laara E, Muir CS. Estimates of the worldwide frequency of sixteen major cancers in 1980. Int J Cancer 1988; 41:184–197.
67 Howley PM. Role of the human papillomaviruses in human cancer. Cancer Res 1991; 51:5019s–5022s.
68 zur Hausen H. Human papillomaviruses in the pathogenesis of anogenital cancer. Virology 1991; 184:9–13.
69 Vousden KH. Regulation of the cell cycle by viral oncoproteins. Semin Cancer Biol 1995; 6:109–116.
70 Moran E. Interaction of adenoviral proteins with pRB and p53. FASEB J 1993; 7:880–885.
71 Vousden KH. Interactions of human papillomavirus transformation proteins with the products of the tumor suppressor genes. FASEB J 1993; 7:872–879.
72 Ludlow JW. Interactions between SV40 large–tumor antigen and the growth suppressor proteins pRb and p53. FASEB J 1993; 7:866–871.
73 Heck DV, Yee CL, Howley PM et al. Efficiency of binding the retinoblastoma protein correlates with the transforming capacity of the E7 oncoproteins of the human papillomaviruses. Proc Natl Acad Sci USA 1992; 89:4442–4446.
74 Sang BC, Barbosa MS. Single amino acid substitutions in "low–risk" human papillomavirus (HPV) type 6 E7 protein enhance features characteristic of the "high–risk" HPV E7 oncoproteins. Proc Natl Acad Sci USA 1992; 89:8063–8067.
75 Lane DP, Crawford LV. T antigen is bound to a host protein in SV40–transformed cells. Nature 1979; 278:261–263.
76 Sarnow P, Ho YS, Williams J et al. Adenovirus E1b–58kd tumor antigen and SV40 large tumor antigen are physically associated with the same 54 kd cellular protein in transformed cells. Cell 1982; 28:387–394.
77 Werness BA, Levine AJ, Howley PM. Association of human papillomavirus types 16 and 18 E6 proteins with p53. Science 1990; 248:76–79.
78 Lechner MS, Mack DH, Finicle AB et al. Human papillomavirus E6 proteins bind p53 in vivo and abrogate p53–mediated repression of transcription. EMBO J 1992; 11:3045–3052.
79 Mietz JA, Unger T, Huibregtse JM et al. The transcriptional transactivation function of wild–type p53 is inhibited by SV40 large T-antigen and by HPV–16 E6 oncoprotein. EMBO J 1992; 11:5013–5020.
80 Yew PR, Berk AJ. Inhibition of p53 transactivation required for transformation by adenovirus early 1B protein. Nature 1992; 357:82–85.
81 Kao CC, Yew PR, Berk AJ. Domains required for in vitro association between the cellular p53 and the adenovirus 2 E1B 55K proteins. Virology 1990; 179:806–814.
82 Yew PR, Liu X, Berk AJ. Adenovirus E1B oncoprotein tethers a transcriptional repression domain to p53. Genes Dev 1994; 8:190–202.
83 Ruppert JM, Stillman B. Analysis of a protein–binding domain of p53. Mol Cell Biol 1993; 13:3811–3820.

84 Bargonetti J, Reynisdottir I, Friedman PN et al. Site–specific binding of wild–type p53 to cellular DNA is inhibited by SV40 T antigen and mutant p53. Genes Dev 1992; 6:1886–1898.
85 Deppert W, Haug M, Steinmayer T. Modulation of p53 protein expression during cellular transformation with simian virus 40. Mol Cell Biol 1987; 7:4453–4463.
86 Lechner MS, Laimins LA. Inhibition of p53 DNA binding by human papillomavirus E6 proteins. J Virol 1994; 68:4264–4273.
87 Marston NJ, Crook T, Vousden KH. Interaction of p53 with MDM2 is independent of E6 and does not mediate wild type transformation suppressor function. Oncogene 1994; 9:2707–2716.
88 Scheffner M, Munger K, Huibregtse JM et al. Targeted degradation of the retinoblastoma protein by human papillomavirus E7–E6 fusion proteins. EMBO J 1992; 11:2425–2431.
89 Kessis TD, Slebos RJ, Nelson WG et al. Human papillomavirus 16 E6 expression disrupts the p53–mediated cellular response to DNA damage. Proc Natl Acad Sci USA 1993; 90:3988–3992.

Chapter 8

Links to Cancer

The first seven chapters of this book gave a detailed treatment of the properties and relationships among the cell cycle regulators—the cyclins, the cyclin–dependent kinases (CDKs), and the cyclin–dependent kinase inhibitors (CKIs). But our primary purpose remains unfulfilled, to formulate a mechanistic explanation of cancer by linking the components of cell cycle machinery to the process of oncogenesis. In this chapter we provide a summation of the wealth of data generated in recent years that implicate various cell cycle regulators as crucial contributors to cancer.

Cyclins

All our knowledge about the cyclins portrays them as positive regulators of the cell cycle; being factors that enhance the kinase activity of CDKs, they are promoters of cellular proliferation. We should therefore speculate that magnifying the effects of one or more cyclins might dispose cells to neoplastic growth. Accordingly, investigators seeking links between cyclins and cancer search for the amplification of cyclin genes or the alteration of the genes such that their products are less amenable to degradation. Either phenomenon would upset the balance between the synthesis and destruction of cyclin protein and result in inappropriately high cyclin levels.

The search for cyclin mutations relevant to cancer has largely been limited to the examination of those proteins whose roles in the cell cycle are well characterized—A, B, D and E. Minimal investigation of the lesser–studied cyclins—C, F, G, H and I—has been reported in the literature.

There is very little data to implicate the A-type and B-type cyclins—the mitotic cyclins—in oncogenesis. In a case of early stage hepatoma, the site of integration for the hepatitis B virus (HBV) genome was found to be the cyclin A gene. The resultant translation product was a chimeric protein in which the N-terminal portion of cyclin A was replaced with an HBV protein. Because the N-terminus of wild–type cyclin A contains the protein's "destruction box" (see chapter 3), the chimeric protein represented a stable, "nondestructible" version of cyclin A. How the altered cyclin might contribute to tumorigenesis is unclear, although the aberrant presence of the protein would likely interfere with checkpoints such as the restriction (R) point and the G1/S transition and thereby deregulate the cell cycle. Besides the single example of this chimeric protein, no correlations between cyclin A and cancer have been noted. Similarly, no firm evidence supporting a role for cyclin B in tumor formation has been reported.

Cyclin E overexpression has been observed in a number of tumor types, particularly breast and colorectal carcinomas.[1-7] But no correlation between cyclin E and a proto–oncogene has been found, and the protein has not been reported to cooperate with known oncogenes in transformation assays, ruling out for the present time a causative role in carcinogenesis.

Among all the cyclins, it is cyclin D1 that is most certainly involved in tumorigenesis. The cyclin D1 gene (also known in the literature as *CCND1*) resides on chromosome 11q13,[8,9] a locus associated with two proto–oncogenes. *PRAD1* was originally identified as a candidate oncogene on 11q13 which was brought adjacent to the parathyroid hormone locus on 11q15 by a chromosomal inversion in a subset of benign parathyroid tumors.[10] The consequence of the rearrangement was the increased expression of *PRAD1* mRNA in the tumor cells.[11] Subsequent analysis has revealed the *PRAD1* gene to be identical to *CCND1*.[12]

A translocation between the 11q13 locus and the immunoglobulin heavy chain locus on chromosome 14 was shown to be a common occurrence in centrocytic B-cell lymphomas.[13,14] The breakpoint on chromosome 11 involved in this translocation, t(11;14)(q13;q32), was believed to fall within a gene that was designated the *BCL1* proto–oncogene. Later studies showed that the *BCL1* breakpoint is almost always most proximal to the *PRAD1* gene, suggesting that the two independently–discovered proto–oncogenes are in fact a single gene encoding cyclin D1.[15-19] Consistent with this premise, the effect of the translocation appears to be the increased expression of *CCND1*, probably resulting from the gene being brought under the control of the immunoglobulin heavy chain enhancer.[18-21] A similar phenomenon has been very infrequently observed in mouse T-cell lymphomas, where the murine *CCND1* homologue is activated by the integration of the murine leukemia virus (MuLV) genome into the chromosome just upstream of the cyclin D1 promoter.[22]

Yet another means by which cyclin D1 appears to participate in oncogenesis is through amplification of the 11q13 locus. Overexpression of cyclin D1 arising by this mechanism correlates with significant proportions of breast, head and neck, esophageal, gastric, and lung carcinomas.[13,23-25] Lastly, in a single breast cancer cell line, increased expression of cyclin D1 has been found to occur as a result of the stabilization of the mRNA due to truncation of the 3' end of the gene.[26]

Cyclin D1 overexpression, irrespective of the chromosomal derangement that causes it, is found in a variety of tumor types. Perhaps of greatest clinical interest is the finding that the incidence of excessive cyclin D1 in breast cancer cases may be one–third or even higher.[1,27-32] Elevated cyclin D1 levels have also been detected in recent studies in cells from esophageal, hepatocellular, colorectal, uterine, head and neck, laryngeal, and both small–cell and non–small–cell carcinomas, parathyroid adenomas, soft tissue sarcomas, malignant melanomas, and various lymphomas.[33-50] Further proof that cyclin D1 can act in a tumorigenic fashion is its ability to cooperate with activated oncogenes in various transformation assays.[51-53] Moreover, transgenic mice in which cyclin D1 is specifically overexpressed in the lymphocytes or mammary cells are more susceptible to neoplastic growth of those cells.[54,55]

The cyclin D2 gene, also known as *CCND2*, resides on chromosome 12p13. Like cyclin D1, the murine homologue of *CCND2* was shown to be activated by MuLV integration in a mouse T-cell lymphoma; the integration site had previously been identified as the *vin–1* proto–oncogene.[56] That the cyclin and the proto–oncogene were identical was confirmed by the overexpression of cyclin D2 in the tumor cells. Yet only a very limited number of alterations of cyclin D2 in tumors have been noted, most often in lymphomas, leukemias and colorectal carcinomas.[2,57,58] Whether the cyclin D3 gene (called *CCND3* and located on chromosome 6p21) has any oncogenic potential remains to be seen, although the data collected thus far suggests that it does not. Why the various D-type cyclins should differ in their relevance to cancer when their cellular functions are rather similar (as far as anyone knows) is unclear.

On the whole, abnormalities in mitotic cyclin expression appear to be rare in cancer, whereas G1 cyclins are altered in a significant number of tumor types. Why should there be such a disparity between the two groups of proteins, since both regulate crucial cell cycle transitions? Part of the answer may lie in studies that have demonstrated that cyclins A and B are required to be degraded in order for the cell to proceed through M phase.[59-61] In the pertinent experiments, nondegradable mutants of cyclins A and B, i.e., truncated versions of the proteins dispossessed of their destruction boxes, induced an M phase arrest when expressed in cells. Presumably, amplified levels of cyclin A or B would similarly delay the completion of mitosis, countering any acceleration of the onset of M phase effected by the aberrant cyclin. It is consequently improbable that the altered expression of mitotic cyclins would contribute significantly to the uncontrolled proliferation of neoplastic cells.

As with cyclins A and B in mitosis, the marked destruction of cyclin E that occurs in S phase (and which is stimulated by cyclin E–CDK2 activity itself; see chapter 6) may be critical for proper execution of the subsequent cell cycle transition, thus leading to a lack of tolerance for deregulated overexpression of cyclin E. In those cases where cyclin E is elevated in tumors, such overexpression may be temporally countered by other regulatory events, such as phosphorylation–mediated disruption of cyclin E–CDK2 complex formation or inhibition of CDK2 activity. In contrast, it does not appear that the deregulation of cyclin D expression has profound negative consequences for cell cycle progression. Perhaps the timed destruction of cyclin D observed in some cells in S phase is able to proceed despite transcriptional deregulation. Alternatively, or in addition, other elements of the varied CDK4 and CDK6 control mechanisms discussed in previous chapters may ensure proper inhibition of activity even when cyclin D is expressed in early G1 phase. With no restriction on the abundance of cyclin D through the entirety of the cell cycle, overexpression of cyclin D could relieve the cell of G1 phase checkpoints as well as interfere with as–of–yet undefined checkpoints in other phases, removing important restraints on cell proliferation. An alternative explanation for the tolerance of cyclin D overexpression is that despite the plethora of controls on cyclin D–CDK4 and –CDK6 function, constitutive activity of this complex is without negative consequence in mammalian cells. Clearly the role of this oncogenically important cyclin in defining the beginnings and ends of cell cycle transitions needs further exploration.

CDKs

If the overexpression of various cyclins can contribute to oncogenesis, then it stands to reason that the alteration of CDKs might contribute as well. As with cyclins, CDKs are thought to act primarily as promoters of cell proliferation. There are two types of mutations that might cause a CDK to inappropriately drive the cell cycle forward: those that cause overexpression of the CDK, e.g., a chromosomal amplification; and those that cause the CDK to assume aberrant kinase activity, e.g., a mutation that converts the protein's Tyr15 residue into a different amino acid that is resistant to phosphorylation.

Because cyclins A and B do not appear to be involved in tumor formation for the reasons enumerated above, it seems unlikely that their mitotic kinase partner, CDK1, is involved either. Bearing out this conjecture, repeated searches for CDK1 amplification in tumor cells have failed to turn up more than a few isolated examples. A mutant form of CDK1 that could propagate kinase activity in the absence of a cyclin partner could in theory accelerate cell proliferation, but no such mutant has ever been found.

The CDK partners of the G1 phase cyclins—CDK2, CDK4 and CDK6—are more credible candidates to be proto–oncogenes. Several derangements of CDK2 have been reported, but not enough to form a coherent pattern that might suggest a role in the development of any particular type of tumor. CDK4 amplifications and mutations have been seen with more consistency in tumor types including gliomas and glioblastomas, sarcomas, and melanomas.[42,50,62-68] Indeed, the recent discovery of a mutant form of CDK4 most strongly supports its candidacy as a human tumor gene. In human melanoma, alteration of Arg24 to Cys renders CDK4 insensitive to the inhibitory effects of p16, presumably allowing unprogrammed activity of the kinase. Further, this mutation appears to cause a familial susceptibility to melanoma, a fact that underscores the importance of this pathway in cell growth deregulation.[67]

CDK6 has not yet been sufficiently investigated to assess its role in tumorigenesis, but given the apparent redundancy in the functions of CDK4 and CDK6 it would not be surprising if the latter were mutated in some proportion of at least a few tumor types. This idea is supported by our own work in human oral cancer cell lines, where CDK6 is consistently hyperactivated and occasionally amplified.[69]

Any oncogenic roles played by CDK3, CDK5, CDK7 or CDK8 have yet to be revealed.

CKIs

Because CKIs act in opposition to cyclins and CDKs, they are viewed as negative regulators of cell proliferation. Further, as explained in chapter 5, all members of the CKI family are considered to be potential tumor suppressor proteins because each one can mediate a cell cycle arrest when overexpressed in particular cell lines. In order to qualify as a bona fide tumor suppressor, however, a protein must satisfy other criteria, the most important of which is that there be a high incidence of alteration of its gene in at least a few tumor types. Because both alleles of a tumor suppressor gene must be inactivated to facilitate tumor formation, we should expect to see homozygous deletion of the gene in cancer cells, hemizygous deletion and mutation of the other allele, or mutation of both alleles.

Derangements of the CIP/KIP inhibitors appear to be rare in cancer. Several published studies and a host of unpublished studies have failed to discern any pattern of deletion or mutation of *p21* in any tumor types.[70-73] While there is data

Table 8.1. Homozygous deletion of *p16* in tumor–derived cell lines

Tumor type	Percentage of cell lines with homozygous deletion of *p16*	References
Astrocytoma	80%–90%	174
Bladder	30%–60%	79,158,174,175
Breast	20%–60%	72,174,176-180
Esophagus	50%–100%	159,181,182
Glioma/Glioblastoma	60%–100%	63,64,147,148,174,183,184
Head and Neck	30%–50%	185-188
Kidney	50%–90%	174,189
Leukemia	20%–100% (varies widely among subtypes)	99,105,107,117,118,151, 154,174,183,190
Lung	0%–70%	121-123,141,173,174,183,191
Lymphoma	0%–50%	112,145,154
Melanoma	20%–70%	50,174,183,192-194
Mesothelioma	80%–90%	195
Neuroblastoma	0%–30%	164,196,197
Osteosarcoma	60%–70%	165,174
Ovary	20%–50%	174,198
Pancreas	20%–70%	162,182,199-201
Prostate	0%	202
Stomach	20%–30%	181

The percentage ranges reported in the table encompass the highest and lowest deletion rates found in the referenced studies with the tumor type, rounded up and down respectively to the nearest multiple of 10%. Studies reporting a deletion rate of less than 5% are rounded down to 0% for the purposes of this table.

to suggest a role for p27 in the development of some leukemias and lymphomas,[58,74] for the most part studies examining the status of *p27* in tumor cells report an absence of alterations.[75-85] This is somewhat surprising in light of the powerful effect of *p27* deletion in the mouse; however, it is interesting to note that the penetrance of pituitary tumorigenesis in different *p27*–knockout founders is rather variable, likely reflecting a significant susceptibility to genetic modifiers. Further, the particular structure affected, the intermediate lobe of the murine pituitary gland, is not present in the human and thus may be peculiarly susceptible to p27 loss in the mouse.[86-88]

At this time, it is not possible to reach any sort of conclusion in regard to p57's role in tumors, but the restricted pattern of p57 expression and its possible role in development and differentiation make it a prime suspect in cancer.

The INK4 inhibitors are more promising candidates to be tumor suppressors. Among all the CKIs, and perhaps all the components of the cell cycle machinery, p16 is the regulator whose gene is most commonly altered in cancer. Homozygous deletion of *p16* alleles has been observed in cell lines derived from tumors in almost every type of human tissue (Table 8.1). Further, *p16* mutations

Table 8.2. Homozygous deletion or mutation of *p16* in primary tumor samples

Tumor Type	Percentage of samples with homozygous deletion (or mutations, when noted) of *p16*	References
Astrocytoma	0%–50%	62,147,203-205
Bladder	0%–30%	152,153,158,175,206-208
Breast	0%	72,177-180,209
Esophagus	0%, 0%–60% (mutations)	210-213
Glioma/Glioblastoma	0%–80% (varies depending on grade of tumor)	62-66,147,148,172, 184,204,205,214-216
Head and Neck	0%; 0-10% (mutations)	186,206,217
Kidney	0%	174,189,206
Leukemia	0%–100% (varies widely among subtypes)	97-119,146,151,218-220
Lung	0%–90% (varies between small–cell and non–small–cell subtypes); 0%–30% (mutations)	46,120-127,206,221
Lymphoma	0%–50%	102,104,112,145,150,161,220
Melanoma	0%–20%; 0%–20% (mutations)	50,192,222-224
Mesothelioma	20%–80%	156,195
Neuroblastoma	0%	81,164,196,197
Osteosarcoma	0%–10%	42,165
Ovary	0%	198,225,226
Pancreas	0-10%; 10%–40% (mutations)	201,227
Prostate	0%	202
Stomach	0%	212,228

The percentage ranges reported in the table encompass the highest and lowest deletion rates found in the referenced studies with the tumor type, rounded up and down respectively to the nearest multiple of 10%. Studies reporting a deletion rate of less than 5% are rounded down to 0% for the purposes of this table.

have been detected in many of the cell lines in which the gene has not been homozygously deleted. Studies of the various mutant forms of p16 found in cancer have revealed that they are impaired in their capacity to suppress cell cycle progression, apparently because of their inability to bind to CDK4 and to inhibit cyclin D–CDK activity.[89-96] One can justifiably stake the claim that a majority of cancerous cell lines have derangements of the p16 gene.

Intriguingly, alterations of *p16* do not seem to be nearly as prevalent in cells from primary tumor samples (Table 8.2). Perhaps the disparity is a reflection of the necessity for p16 inactivation in the immortalization of cell lines, whereas tumor formation in living human tissues does not require the elimination of p16 function because it can proceed by alternative mechanisms. Remarkably, defective p16 appears to be a hallmark in certain tumor types yet absent in other closely related tumor types. Homozygous *p16* deletion is so prevalent in cases of acute lymphoblastic leukemia of T–cell lineage (T–ALL) that some investigators consider it to be a critical event in the pathogenesis of that particular type of cancer.[97-119] Yet *p16* alteration is less commonly observed in acute lymphoblastic leukemia of B–cell lineage (B–ALL), and it is considerably rarer in other leukemia types. Another intriguing contrast is that there is a high incidence of *p16* deletion in non–small–cell lung carcinomas (NSCLCs), particularly in late–stage cases, but a very low incidence in small–cell lung carcinomas (SCLCs).[46,120-127]

The p16 gene resides on chromosome 9p21, where the locus for familial predisposition to melanoma (designated *MLM*) has been mapped. Supporting the premise that *p16* is identical to *MLM*, mutations in the p16 gene have been shown to be present in a large proportion of individuals in a number of melanoma kindreds.[128-135] It is therefore tempting to surmise that *p16* alteration is an essential event for the formation of melanomas, but other evidence surrounding this proposition is dubious at best. *p16* is homozygously deleted in a substantial percentage of melanoma–derived cell lines (Table 8.1), but as we noted above this observation does not necessarily reflect on the relevance of p16 to tumorigenesis in vivo. No study has found *p16* deletion in more than 20% of primary melanoma samples, nor has any study found mutation of the gene in more than 20% of samples (Table 8.2), suggesting that p16 is a significant but not crucial contributor to melanoma formation.

Yet the infrequent occurrence of *p16* deletions and mutations in most types of primary tumor samples (including melanomas) does not necessarily rule out a causative role of p16 in the formation of those cancers. A number of recent studies have demonstrated that the expression of p16 protein can be silenced by the methylation of cytosine residues in the promoter region and exon 1 of the *p16* gene. Indeed, a significant proportion of samples from a number of tumor types display aberrant *p16* methylation and, correspondingly, the absence of p16 protein; among these tumor types are bladder, breast, prostate, kidney and colon, in which *p16* deletion is rare.[136-143] The possibility exists that methylation constitutes as important (if not more important) a source of *p16* oncogenicity as deletions and mutations, although it is not yet clear by what mechanism this methylation occurs.

The extent to which the p15 inhibitor, which bears striking structural similarity to p16, plays a role in carcinogenesis is far from clear. Complicating any attempts to correlate alterations in the p15 gene to incidence of cancer is the fact that *p15* resides on chromosome 9p21 a mere 25 kilobases apart from *p16*. The p15 gene is homozygously deleted in a wide variety of tumor types, but the significance of this phenomenon is brought into question by the observation that *p15* is usually found to be codeleted with *p16*; cases in which *p15* is altered and *p16* is spared are rarer than cases in which the opposite is true.[66,72,103,106,107,109,113,115,123-125,127,144-163] A host of questions arise. Does a cancer cell's need to abolish p16 expression cause it to inadvertently eliminate p15 as well, or does the silencing of p15 expression in and of itself contribute to the cell's pathological state? Is p15 a bona fide tumor suppressor or a pretender, masked by its gene's proximity to that of p16? Since the functional roles of p15 and p16 seem to be at least partially redundant, perhaps both need to be eliminated for a cancer cell's full hyperproliferative capability to be unleashed, whereas the elimination of one allows tumor formation to proceed only at a slower pace. Or perhaps certain cell types require both the inhibitors' genes to be silenced to become tumorigenic, whereas other cell types are not so demanding. Further investigation will be required to elucidate the relationship of p15 and p16 as it pertains to cancer.

In contrast to its close relatives, p18 is almost certainly not a tumor suppressor, judging from the negative results of a number of studies probing for altered *p18* in tumor types in which *p15* and *p16* mutations have been observed.[81,113,116,124,127,149,151,163-167] There has been minimal investigation to date concerning the status of p19 in tumor cells; whether the inhibitor will resemble p15 and p16 on the one hand or p18 on the other remains to be seen.

Other Cell Cycle–Related Proteins

As described in previous chapters, the retinoblastoma susceptibility gene product (pRb) and the p53 protein are well–characterized tumor suppressor proteins that have direct involvement in cell cycle regulation. Mutations in the pRb gene, *RB1*, occur in a wide array of tumors including retinoblastomas, osteosarcomas, mesenchymal tumors, small cell lung carcinomas, breast carcinomas, bladder carcinomas, synovial sarcomas, prostate carcinomas and leukemias (chapter 6). *p53* is altered in so many different tumor types that it has earned the distinction of being the most commonly mutated gene in cancer; some estimates of the incidence of *p53* mutation in tumors surpass 50% (chapter 7). Any discussion of the relationship of the cell cycle to cancer would be incomplete without mention of these two enormously important genes.

Recently *CDC25* has been implicated as a possible proto–oncogene; all three isoforms of CDC25 possess the ability to cooperate with an activated Ras oncoprotein in transformation assays, and CDC25B is overexpressed in a significant proportion of breast cancer cases.[168] This finding is consistent with CDC25's known cellular function, to dephosphorylate and thereby activate cyclin–CDK complexes (chapter 4). An aberrantly high quantity of CDC25 would presumably overwhelm Wee1's ability to inhibit cyclin–CDKs via phosphorylation, thus circumventing important checkpoint controls and facilitating cell proliferation.

Lastly, we should consider the relationships of confirmed proto–oncogenes to the cell cycle. Raf1, for instance, is believed to interact with and activate

CDC25 and Myc can transcriptionally transactivate the CDC25 gene;[169,170] constitutively active mutants of the two oncoproteins would generate an omnipresent pool of active CDC25 that would in turn induce cyclin–CDK activity at inappropriate times. It seems likely that many, if not all, oncogene products interfere with the core cell cycle machinery in some indirect fashion and thereby contribute to tumorigenesis.

A Central Pathway for Cell Cycle Deregulation?

Let us compile a list of the cell cycle regulators which we can correlate with the incidence of cancer with some degree of confidence: cyclin D1, which we know to be the *PRAD1* oncogene, and perhaps cyclins D2 and D3 and E; CDK4, which is amplified in a number of tumors, and possibly CDK6; p16 and p15, putative tumor suppressor genes which are deleted or mutated in many types of cancers. Remarkably, most of these proteins constitute a single regulatory pathway centered upon the cyclin D–CDK complex (Fig. 8.1). It should not escape our notice that the primary downstream target of this complex is pRb, and that this complex can be indirectly inhibited by p53 via p21. Moreover, if CDC25 or some closely related protein is responsible for activating CDK4 (such a role on the part of CDC25 has already been demonstrated for CDK1 and CDK2), we can connect CDC25 and its oncogenic upstream regulators, Raf1 and Myc, to the cyclin D–CDK pathway as well. The prevalence of mutations in this pathway in cancer is so striking that it has been suggested that the pathway is disrupted in *all* cases of human cancer.[171]

The veracity of this conjecture is borne out by recent comprehensive studies of two tumor types. Among 120 human glioblastoma cases, 40% lacked both alleles of *p16*, 14% lacked both alleles of *RB1*, and 12% exhibited amplified CDK4; with only two exceptions, each tumor had but one of these abnormalities. A further 30% had lost one allele of the p16 and/or pRb genes. In just 6% of the cases were none of the studied cell cycle regulators altered.[172] Out of 55 SCLCs, 48 had deficient or mutant pRb but possessed detectable wild–type p16. Six of the remaining seven tumors lacked p16 protein but expressed wild–type pRb at normal levels.[173]

Why would this particular pathway be so crucial to the incidence of cancer? There is an emerging body of evidence suggesting that the phosphorylation of pRb by cyclin D–CDK complexes is closely associated with the cell's passage through the R point, after which the cell is committed to completing a round of DNA replication and the subsequent mitosis (see chapter 3). The R point is arguably the most important checkpoint in the entirety of the mammalian cell cycle, for it is quite literally the point of no return. If we assume that the cyclin

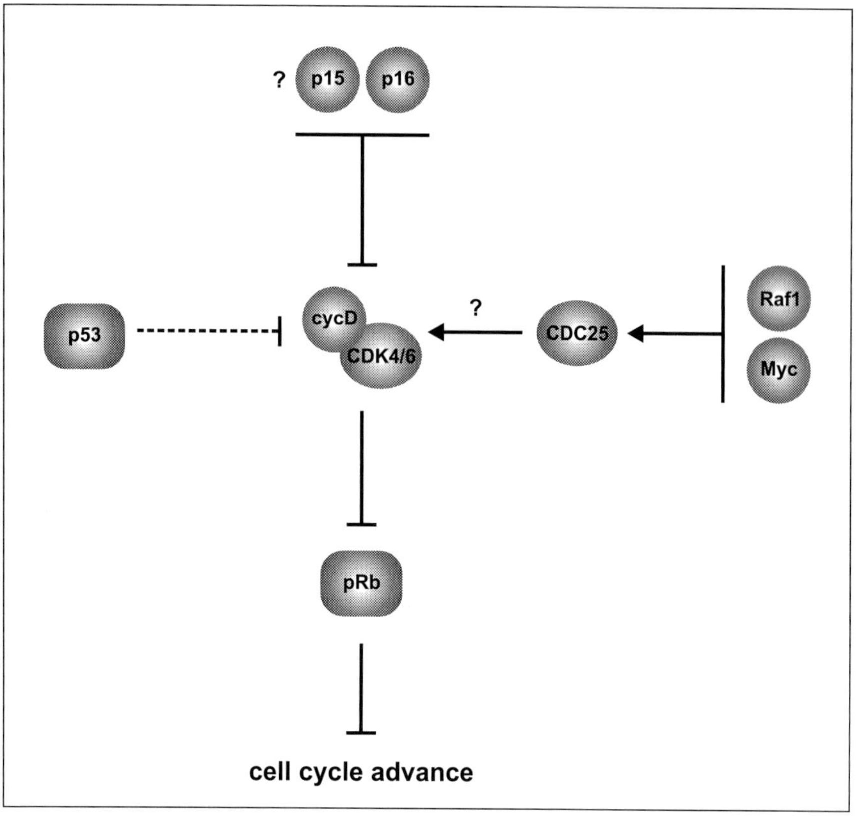

Fig. 8.1. Diagram depicting the cyclin D-CDK pathway (also known as the pRb pathway), in which a number of putative oncoproteins and tumor suppressors participate. It has been proposed that this pathway is altered in some way in all cases of human cancer. Arrows indicate an activation or upregulation of the target protein, and bars indicate an inhibition or downregulation of the target protein; solid lines indicate a direct interaction between the proteins, and dotted lines indicate an indirect interaction. The question marks indicate that it is not clear whether p15 is a bona fide tumor suppressor protein, and that CDC25 (or a closely related protein) is a suspected but not proven regulator of cyclin D-CDK complexes.

D–CDK pathway is indeed linked to the R point, then we might judge this pathway to be the most crucial in cell cycle regulation. An appropriate mutation in any of the elements involved in the pathway will lead to inactivation of pRb, stimulating the cell to indiscriminately traverse the R point. This deregulation of the R point would be a significant step—perhaps an essential step—towards neoplastic growth.

The apparent exclusivity of mutations in the cyclin D-CDK pathway, also known as the pRb pathway, has been cited as evidence that p16, cyclin D1, CDK4 and their associated regulators serve solely or predominantly to regulate pRb, and thus a single hit in any element of this pathway is essentially equivalent to any other. Loss of p16 and pRb in the same tumor, for example, would be redundant and thus not selected. However, it should also be noted that each of the "hits" is arguably pleiotropic—activation of CDK4 or loss of pRb is likely to alter function or expression of myriad other cellular proteins, and the spectrum of these changes may depend on cell type. It is highly suggestive that mutations of individual pathway members appear to be nonrandom within a given tumor type—this is reflected by the asymmetry in the number of *p16* and *RB1* deletions noted in glioblastomas and SCLCs in the studies described above. Perhaps CDK4 and CDK6 have targets other than pRb (and the other pocket proteins) that contribute to the transformed phenotype more highly in some cell types than others; in the same vein, predominant loss of pRb in some tumors may occur because the activity of its regulatory kinases is under such stringent control that loss of a single inhibitor or overexpression of a cyclin or CDK subunit is not sufficient to deregulate activity. Clearly, the work in the field to date has rapidly defined a critical regulatory pathway in proliferation and cancer, yet the future holds many important unknowns that may be crucial to our understanding of growth deregulation in specific tumors.

References

1 Buckley MF, Sweeney KJ, Hamilton JA et al. Expression and amplification of cyclin genes in human breast cancer. Oncogene 1993; 8:2127–2133.
2 Leach FS, Elledge SJ, Sherr CJ et al. Amplification of cyclin genes in colorectal carcinomas. Cancer Res 1993; 53:1986–1989.
3 Keyomarsi K, Pardee AB. Redundant cyclin overexpression and gene amplification in breast cancer cells. Proc Natl Acad Sci USA 1993; 90:1112–1116.
4 Keyomarsi K, O'Leary N, Molnar G et al. Cyclin E, a potential prognostic marker for breast cancer. Cancer Res 1994; 54:380–385.
5 Akama Y, Yasui W, Yokozaki H et al. Frequent amplification of the cyclin E gene in human gastric carcinomas. Jpn J Cancer Res 1995; 86:617–621.
6 Kitahara K, Yasui W, Kuniyasu H et al. Concurrent amplification of cyclin E and CDK2 genes in colorectal carcinomas. Int J Cancer 1995; 62:25–28.
7 Wang A, Yoshimi N, Suzui M et al. Different expression patterns of cyclins A, D1 and E in human colorectal cancer. J Cancer Res Clin Oncol 1996; 122:122–126.
8 Inaba T, Matsushime H, Valentine M et al. Genomic organization, chromosomal localization, and independent expression of human cyclin D genes. Genomics 1992; 13:565–574.
9 Xiong Y, Menninger J, Beach D et al. Molecular cloning and chromosomal mapping of CCND genes encoding human D-type cyclins. Genomics 1992; 13:575–584.
10 Arnold A, Kim HG, Gaz RD et al. Molecular cloning and chromosomal mapping of DNA rearranged with the parathyroid hormone gene in a parathyroid adenoma. J Clin Invest 1989; 83:2034–2040.

11 Rosenberg CL, Kim HG, Shows TB et al. Rearrangement and overexpression of D11S287E, a candidate oncogene on chromosome 11q13 in benign parathyroid tumors. Oncogene 1991; 6:449–453.
12 Motokura T, Bloom T, Kim HG et al. A novel cyclin encoded by a bcl1–linked candidate oncogene. Nature 1991; 350:512–515.
13 Lammie GA, Peters G. Chromosome 11q13 abnormalities in human cancer. Cancer Cells 1991; 3:413–420.
14 Raffeld M, Sander CA, Yano T et al. Mantle cell lymphoma: an update. Leuk Lymphoma 1992; 8:161–166.
15 Withers DA, Harvey RC, Faust JB et al. Characterization of a candidate bcl-1 gene. Mol Cell Biol 1991; 11:4846–4853.
16 Brookes S, Lammie GA, Schuuring E et al. Linkage map of a region of human chromosome band 11q13 amplified in breast and squamous cell tumors. Genes Chromosom Cancer 1992; 4:290–301.
17 Williams ME, Swerdlow SH, Rosenberg CL et al. Characterization of chromosome 11 translocation breakpoints at the bcl–1 and PRAD1 loci in centrocytic lymphoma. Cancer Res 1992; 52:5541s-5544s.
18 de Boer CJ, Loyson S, Kluin PM et al. Multiple breakpoints within the BCL–1 locus in B–cell lymphoma: rearrangements of the cyclin D1 gene. Cancer Res 1993; 53:4148–4152.
19 Raynaud SD, Bekri S, Leroux D et al. Expanded range of 11q13 breakpoints with differing patterns of cyclin D1 expression in B–cell malignancies. Genes Chromosom Cancer 1993; 8:80–87.
20 Rosenberg CL, Wong E, Petty EM et al. PRAD1, a candidate BCL1 oncogene: mapping and expression in centrocytic lymphoma. Proc Natl Acad Sci USA 1991; 88:9638–9642.
21 Seto M, Yamamoto K, Iida S et al. Gene rearrangement and overexpression of PRAD1 in lymphoid malignancy with t(11;14)(q13;q32) translocation. Oncogene 1992; 7:1401–1406.
22 Lammie GA, Smith R, Silver J et al. Proviral insertions near cyclin D1 in mouse lymphomas: a parallel for BCL1 translocations in human B-cell neoplasms. Oncogene 1992; 7:2381–2387.
23 Jiang W, Zhang YJ, Kahn SM et al. Altered expression of the cyclin D1 and retinoblastoma genes in human esophageal cancer. Proc Natl Acad Sci USA 1993; 90:9026–9030.
24 Tsuruta H, Sakamoto H, Onda M et al. Amplification and overexpression of EXP1 and EXP2/Cyclin D1 genes in human esophageal carcinomas. Biochem Biophys Res Commun 1993; 196:1529–1536.
25 Fantl V, Smith R, Brookes S et al. Chromosome 11q13 abnormalities in human breast cancer. Cancer Surv 1993; 18:77-94.
26 Lebwohl DE, Muise–Helmericks R, Sepp–Lorenzino L et al. A truncated cyclin D1 gene encodes a stable mRNA in a human breast cancer cell line. Oncogene 1994; 9:1925-1929.
27 Bartkova J, Lukas J, Muller H et al. Cyclin D1 protein expression and function in human breast cancer. Int J Cancer 1994; 57:353–361.
28 Gillett C, Fantl V, Smith R et al. Amplification and overexpression of cyclin D1 in breast cancer detected by immunohistochemical staining. Cancer Res 1994; 54:1812–1817.
29 Zhang SY, Caamano J, Cooper F et al. Immunohistochemistry of cyclin D1 in human breast cancer. Am J Clin Pathol 1994; 102:695–698.
30 Dickson C, Fantl V, Gillett C et al. Amplification of chromosome band 11q13 and a role for cyclin D1 in human breast cancer. Cancer Lett 1995; 90:43–50.
31 McIntosh GG, Anderson JJ, Milton I et al. Determination of the prognostic value of cyclin D1 overexpression in breast cancer. Oncogene 1995; 11:885–891.
32 Zukerberg LR, Yang WI, Gadd M et al. Cyclin D1 (PRAD1) protein expression in breast cancer: approximately one–third of infiltrating mammary carcinomas show overexpression of the cyclin D1 oncogene. Mod Path 1995; 8:560–567.
33 Yoshida T, Sakamoto H, Terada M. Amplified genes in cancer in upper digestive tract. Semin Cancer Biol 1993; 4:33–40.
34 Zhang YJ, Jiang W, Chen CJ et al. Amplification and overexpression of cyclin D1 in human hepatocellular carcinoma. Biochem Biophys Res Commun 1993; 196:1010–1016.
35 Bartkova J, Lukas J, Strauss M et al. The PRAD–1/cyclin D1 oncogene product accumulates aberrantly in a subset of colorectal carcinomas. Int J Cancer 1994; 58:568–573.

36 Bosch F, Jares P, Campo E et al. PRAD–1/cyclin D1 gene overexpression in chronic lymphoproliferative disorders: a highly specific marker of mantle cell lymphoma. Blood 1994; 84:2726–2732.
37 Callender T, el–Naggar AK, Lee MS et al. PRAD-1 (CCND1)/cyclin D1 oncogene amplification in primary head and neck squamous cell carcinoma. Cancer 1994; 74:152–158.
38 Jares P, Fernandez PL, Campo E et al. PRAD–1/cyclin D1 gene amplification correlates with messenger RNA overexpression and tumor progression in human laryngeal carcinomas. Cancer Res 1994; 54:4813–4817.
39 Nishida N, Fukuda Y, Komeda T et al. Amplification and overexpression of the cyclin D1 gene in aggressive human hepatocellular carcinoma. Cancer Res 1994; 54:3107–3110.
40 Bartkova J, Lukas J, Strauss M et al. Cyclin D1 oncoprotein aberrantly accumulates in malignancies of diverse histogenesis. Oncogene 1995; 10:775–778.
41 Delmer A, Ajchenbaum–Cymbalista F, Tang R et al. Over–expression of cyclin D1 in chronic B–cell malignancies with abnormality of chromosome 11q13. Br J Haematol 1995; 89:798–804.
42 Maelandsmo GM, Berner JM, Florenes VA et al. Homozygous deletion frequency and expression levels of the CDKN2 gene in human sarcomas—relationship to amplification and mRNA levels of CDK4 and CCND1. Br J Cancer 1995; 72:393–398.
43 Michalides R, van Veelen N, Hart A et al. Overexpression of cyclin D1 correlates with recurrence in a group of forty–seven operable squamous cell carcinomas of the head and neck. Cancer Res 1995; 55:975–978.
44 Naitoh H, Shibata J, Kawaguchi A et al. Overexpression and localization of cyclin D1 mRNA and antigen in esophageal cancer. Am J Pathol 1995; 146:1161–1169.
45 Nakagawa H, Zukerberg L, Togawa K et al. Human cyclin D1 oncogene and esophageal squamous cell carcinoma. Cancer 1995; 76:541–549.
46 Shapiro GI, Edwards CD, Kobzik L et al. Reciprocal Rb inactivation and p16[INK4] expression in primary lung cancers and cell lines. Cancer Res 1995; 55:505–509.
47 Swerdlow SH, Yang WI, Zukerberg LR et al. Expression of cyclin D1 protein in centrocytic/mantle cell lymphomas with and without rearrangement of the BCL1/cyclin D1 gene. Hum Pathol 1995; 26:999–1004.
48 Zukerberg LR, Yang WI, Arnold A et al. Cyclin D1 expression in non–Hodgkin's lymphomas. Detection by immunohistochemistry. Am J Clin Pathol 1995; 103:756–760.
49 Betticher DC, Heighway J, Hasleton PS et al. Prognostic significance of CCND1 (cyclin D1) overexpression in primary resected non–small–cell lung cancer. Br J Cancer 1996; 73:294–300.
50 Maelandsmo GM, Florenes VA, Hovig E et al. Involvement of the pRb/p16/cdk4/cyclin D1 pathway in the tumorigenesis of sporadic malignant melanomas. Br J Cancer 1996; 73:909–916.
51 Jiang W, Kahn SM, Zhou P et al. Overexpression of cyclin D1 in rat fibroblasts causes abnormalities in growth control, cell cycle progression and gene expression. Oncogene 1993; 8:3447–3457.
52 Hinds PW, Dowdy SF, Eaton EN et al. Function of a human cyclin gene as an oncogene. Proc Natl Acad Sci USA 1994; 91:709–713.
53 Lovec H, Grzeschiczek A, Kowalski MB et al. Cyclin D1/bcl-1 cooperates with myc genes in the generation of B-cell lymphoma in transgenic mice. EMBO J 1994; 13:3487–3495.
54 Bodrug SE, Warner BJ, Bath ML et al. Cyclin D1 transgene impedes lymphocyte maturation and collaborates in lymphomagenesis with the myc gene. EMBO J 1994; 13:2124–2130.
55 Wang TC, Cardiff RD, Zukerberg L et al. Mammary hyperplasia and carcinoma in MMTV–cyclin D1 transgenic mice. Nature 1994; 369:669–671.
56 Hanna Z, Jankowski M, Tremblay P et al. The Vin-1 gene, identified by provirus insertional mutagenesis, is the cyclin D2. Oncogene 1993; 8:1661–1666.
57 Delmer A, Ajchenbaum–Cymbalista F, Tang R et al. Overexpression of cyclin D2 in chronic B-cell malignancies. Blood 1995; 85:2870–2876.
58 Hoglund M, Johansson B, Pedersen–Bjergaard J et al. Molecular characterization of 12p abnormalities in hematologic malignancies: deletion of KIP1, rearrangement of TEL, and amplification of CCND2. Blood 1996; 87:324–330.
59 Murray AW, Solomon MJ, Kirschner MW. The role of cyclin synthesis and degradation in the control of maturation promoting factor activity. Nature 1989; 339:280–286.

60 Luca FC, Shibuya EK, Dohrmann CE et al. Both cyclin A delta 60 and B delta 97 are stable and arrest cells in M-phase, but only cyclin B delta 97 turns on cyclin destruction. EMBO J 1991; 10:4311–4320.
61 Gallant P, Nigg EA. Cyclin B2 undergoes cell cycle–dependent nuclear translocation and, when expressed as a non–destructible mutant, causes mitotic arrest in HeLa cells. J Cell Biol 1992; 117:213–224.
62 Schmidt EE, Ichimura K, Reifenberger G et al. CDKN2 (p16/MTS1) gene deletion or CDK4 amplification occurs in the majority of glioblastomas. Cancer Res 1994; 54:6321–6324.
63 He J, Allen JR, Collins VP et al. CDK4 amplification is an alternative mechanism to p16 gene homozygous deletion in glioma cell lines. Cancer Res 1994; 54:5804–5807.
64 He J, Olson JJ, James CD. Lack of $p16^{INK4}$ or retinoblastoma protein (pRb), or amplification–associated overexpression of cdk4 is observed in distinct subsets of malignant glial tumors and cell lines. Cancer Res 1995; 55:4833–4836.
65 Nishikawa R, Furnari FB, Lin H et al. Loss of $P16^{INK4}$ expression is frequent in high grade gliomas. Cancer Res 1995; 55:1941–1945.
66 Sonoda Y, Yoshimoto T, Sekiya T. Homozygous deletion of the MTS1/p16 and MTS2/p15 genes and amplification of the CDK4 gene in glioma. Oncogene 1995; 11:2145–2149.
67 Wolfel T, Hauer M, Schneider J et al. A $p16^{INK4a}$–insensitive CDK4 mutant targeted by cytolytic T lymphocytes in a human melanoma. Science 1995; 269:1281–1284.
68 Zuo L, Weger J, Yang Q et al. Germline mutations in the $p16^{INK4a}$ binding domain of CDK4 in familial melanoma. Nat Genet 1996; 12:97–99.
69 Timmerman S, Hinds PW, Munger K. Highly elevated activity of the pRB–kinase cdk6 in oral epithelial carcinoma cells. Cell Growth Differ 1997; in press.
70 Shiohara M, el–Deiry WS, Wada M et al. Absence of WAF1 mutations in a variety of human malignancies. Blood 1994; 84:3781–3784.
71 Li YJ, Laurent–Puig P, Salmon RJ et al. Polymorphisms and probable lack of mutation in the WAF1–CIP1 gene in colorectal cancer. Oncogene 1995; 10:599–601.
72 Musgrove EA, Lilischkis R, Cornish AL et al. Expression of the cyclin–dependent kinase inhibitors $p16^{INK4}$, $p15^{INK4B}$ and $p21^{WAF1/CIP1}$ in human breast cancer. Int J Cancer 1995; 63:584–591.
73 Sun Y, Hildesheim A, Li H et al. No point mutation but a codon 31ser→arg polymorphism of the WAF–1/CIP–1/p21 tumor suppressor gene in nasopharyngeal carcinoma (NPC): the polymorphism distinguishes Caucasians from Chinese. Cancer Epidemiol Biomarkers Prev 1995; 4:261–267.
74 Morosetti R, Kawamata N, Gombart AF et al. Alterations of the $p27^{KIP1}$ gene in non–Hodgkin's lymphomas and adult T-cell leukemia/lymphoma. Blood 1995; 86:1924–1930.
75 Cave H, Gerard B, Martin E et al. Loss of heterozygosity in the chromosomal region 12p12–13 is very common in childhood acute lymphoblastic leukemia and permits the precise localization of a tumor–suppressor gene distinct from $p27^{KIP1}$. Blood 1995; 86:3869–3875.
76 Kawamata N, Morosetti R, Miller CW et al. Molecular analysis of the cyclin–dependent kinase inhibitor gene p27/Kip1 in human malignancies. Cancer Res 1995; 55:2266–2269.
77 Pietenpol JA, Bohlander SK, Sato Y et al. Assignment of the human $p27^{Kip1}$ gene to 12p13 and its analysis in leukemias. Cancer Res 1995; 55:1206–1210.
78 Ponce–Castaneda MV, Lee MH, Latres E et al. $p27^{Kip1}$: chromosomal mapping to 12p12–12p13.1 and absence of mutations in human tumors. Cancer Res 1995; 55:1211–1214.
79 Southgate J, Proffitt J, Roberts P et al. Loss of cyclin–dependent kinase inhibitor genes and chromosome 9 karyotypic abnormalities in human bladder cancer cell lines. Br J Cancer 1995; 72:1214–1218.
80 Ferrando AA, Balbin M, Pendas AM et al. Mutational analysis of the human cyclin–dependent kinase inhibitor $p27^{kip1}$ in primary breast carcinomas. Hum Genet 1996; 97:91–94.
81 Kawamata N, Seriu T, Koeffler HP et al. Molecular analysis of the cyclin–dependent kinase inhibitor family: p16(CDKN2/MTS1/INK4A), p18(INK4C) and p27(Kip1) genes in neuroblastomas. Cancer 1996; 77:570–575.
82 Orlow I, Iavarone A, Crider–Miller SJ et al. Cyclin–dependent kinase inhibitor $p57^{KIP2}$ in soft tissue sarcomas and Wilms' tumors. Cancer Res 1996; 56:1219–1221.
83 Spirin KS, Simpson JF, Takeuchi S et al. p27/Kip1 mutation found in breast cancer. Cancer Res 1996; 56:2400–2404.

84 Stegmaier K, Takeuchi S, Golub TR et al. Mutational analysis of the candidate tumor suppressor genes TEL and KIP1 in childhood acute lymphoblastic leukemia. Cancer Res 1996; 56:1413–1417.
85 Takeuchi S, Mori N, Koike M et al. Frequent loss of heterozygosity in region of the KIP1 locus in non–small cell lung cancer: evidence for a new tumor suppressor gene on the short arm of chromosome 12. Cancer Res 1996; 56:738–740.
86 Nakayama K, Ishida N, Shirane M et al. Mice lacking p27(Kip1) display increased body size, multiple organ hyperplasia, retinal dysplasia, and pituitary tumors. Cell 1996; 85:707–720.
87 Kiyokawa H, Kineman RD, Manova–Todorova KO et al. Enhanced growth of mice lacking the cyclin–dependent kinase inhibitor function of p27(Kip1). Cell 1996; 85:721–732.
88 Fero ML, Rivkin M, Tasch M et al. A syndrome of multiorgan hyperplasia with features of gigantism, tumorigenesis, and female sterility in p27(Kip1)–deficient mice. Cell 1996; 85:733–744.
89 Koh J, Enders GH, Dynlacht BD et al. Tumour–derived p16 alleles encoding proteins defective in cell–cycle inhibition. Nature 1995; 375:506–510.
90 Ranade K, Hussussian CJ, Sikorski RS et al. Mutations associated with familial melanoma impair p16^{INK4} function. Nat Genet 1995; 10:114–116.
91 Reymond A, Brent R. p16 proteins from melanoma–prone families are deficient in binding to Cdk4. Oncogene 1995; 11:1173–1178.
92 Wick ST, Dubay MM, Imanil I et al. Biochemical and mutagenic analysis of the melanoma tumor suppressor gene product/p16. Oncogene 1995; 11:2013–2019.
93 Yang R, Gombart AF, Serrano M et al. Mutational effects on the p16^{INK4a} tumor suppressor protein. Cancer Res 1995; 55:2503–2506.
94 Lilischkis R, Sarcevic B, Kennedy C et al. Cancer–associated mis–sense and deletion mutations impair p16^{INK4} CDK inhibitory activity. Int J Cancer 1996; 66:249–254.
95 Parry D, Peters G. Temperature–sensitive mutants of p16^{CDKN2} associated with familial melanoma. Mol Cell Biol 1996; 16:3844–3852.
96 Yang R, Serrano M, Slater J et al. Analysis of p16^{INK4a} and its interaction with CDK4. Biochem Biophys Res Commun 1996; 218:254–259.
97 Duro D, Flexor MA, Bernard O et al. Alterations of the putative tumor suppressor gene p16/MTS1 in human hematological malignancies. C R Acad Sci III 1994; 317:913–919.
98 Hebert J, Cayuela JM, Berkeley J et al. Candidate tumor–suppressor genes MTS1 (p16^{INK4A}) and MTS2 (p15^{INK4B}) display frequent homozygous deletions in primary cells from T– but not from B–cell lineage acute lymphoblastic leukemias. Blood 1994; 84:4038–4044.
99 Ogawa S, Hirano N, Sato N et al. Homozygous loss of the cyclin–dependent kinase 4–inhibitor (p16) gene in human leukemias. Blood 1994; 84:2431–2435.
100 Cayuela JM, Hebert J, Sigaux F. Homozygous MTS1 (p16^{INK4A}) deletion in primary tumor cells of 163 leukemic patients. Blood 1995; 85:854.
101 Delmer A, Tang R, Senamaud–Beaufort C et al. Alterations of cyclin–dependent kinase 4 inhibitor (p16$^{INK4A/MTS1}$) gene structure and expression in acute lymphoblastic leukemias. Leukemia 1995; 9:1240–1245.
102 Fizzotti M, Cimino G, Pisegna S et al. Detection of homozygous deletions of the cyclin–dependent kinase 4 inhibitor (p16) gene in acute lymphoblastic leukemia and association with adverse prognostic features. Blood 1995; 85:2685–2690.
103 Haidar MA, Cao XB, Manshouri T et al. p16^{INK4A} and p15^{INK4B} gene deletions in primary leukemias. Blood 1995; 86:311–315.
104 Ogawa S, Hangaishi A, Miyawaki S et al. Loss of the cyclin–dependent kinase 4-inhibitor (p16; MTS1) gene is frequent in and highly specific to lymphoid tumors in primary human hematopoietic malignancies. Blood 1995; 86:1548–1556.
105 Ohnishi H, Kawamura M, Ida K et al. Homozygous deletions of p16/MTS1 gene are frequent but mutations are infrequent in childhood T–cell acute lymphoblastic leukemia. Blood 1995; 86:1269–1275.
106 Okuda T, Shurtleff SA, Valentine MB et al. Frequent deletion of p16$^{INK4a/MTS1}$ and p15$^{INK4b/MTS2}$ in pediatric acute lymphoblastic leukemia. Blood 1995; 85:2321–2330.
107 Otsuki T, Clark HM, Wellmann A et al. Involvement of CDKN2 (p16$^{INK4A/MTS1}$) and p15$^{INK4B/MTS2}$ in human leukemias and lymphomas. Cancer Res 1995; 55:1436–1440.

108 Quesnel B, Preudhomme C, Philippe N et al. p16 gene homozygous deletions in acute lymphoblastic leukemia. Blood 1995; 85:657–663.
109 Rasool O, Heyman M, Brandter LB et al. p15[ink4B] and p16[ink4] gene inactivation in acute lymphocytic leukemia. Blood 1995; 85:3431–3436.
110 Schroder M, Mathieu U, Dreyling MH et al. CDKN2 gene deletion is not found in chronic lymphoid leukaemias of B– and T–cell origin but is frequent in acute lymphoblastic leukaemia. Br J Haematol 1995; 91:865–870.
111 Sill H, Goldman JM, Cross NC. Homozygous deletions of the p16 tumor–suppressor gene are associated with lymphoid transformation of chronic myeloid leukemia. Blood 1995; 85:2013–2016.
112 Stranks G, Height SE, Mitchell P et al. Deletions and rearrangement of CDKN2 in lymphoid malignancy. Blood 1995; 85:893–901.
113 Takeuchi S, Bartram CR, Seriu T et al. Analysis of a family of cyclin–dependent kinase inhibitors: p15/MTS2/INK4B, p16/MTS1/INK4A, and p18 genes in acute lymphoblastic leukemia of childhood. Blood 1995; 86:755–760.
114 Cayuela JM, Madani A, Sanhes L et al. Multiple tumor–suppressor gene 1 inactivation is the most frequent genetic alteration in T-cell acute lymphoblastic leukemia. Blood 1996; 87:2180–2186.
115 Heyman M, Rasool O, Borgonovo Brandter L et al. Prognostic importance of p15[INK4B] and p16[INK4] gene inactivation in childhood acute lymphocytic leukemia. J Clin Oncol 1996; 14:1512–1520.
116 Iolascon A, Faienza MF, Coppola B et al. Homozygous deletions of cyclin–dependent kinase inhibitor genes, p16(INK4A) and p18, in childhood T cell lineage acute lymphoblastic leukemias. Leukemia 1996; 10:255–260.
117 Jagasia AA, Sher DA, Le Moine PJ et al. Deletion or lack of expression of CDKN2 (CDK4I/MTS1/INK4A) and MTS2 (INK4B) in acute lymphoblastic leukemia cell lines reflects the phenotype of the uncultured primary leukemia cells. Leukemia 1996; 10:624–628.
118 Kees UR, Ranford PR, Hatzis M. Deletions of the p16 gene in pediatric leukemia and corresponding cell lines. Oncogene 1996; 12:2235–2239.
119 Nakao M, Yokota S, Kaneko H et al. Alterations of CDKN2 gene structure in childhood acute lymphoblastic leukemia: mutations of CDKN2 are observed preferentially in T lineage. Leukemia 1996; 10:249–254.
120 Hayashi N, Sugimoto Y, Tsuchiya E et al. Somatic mutations of the MTS (multiple tumor suppressor) 1/CDK4l (cyclin–dependent kinase–4 inhibitor) gene in human primary non–small cell lung carcinomas. Biochem Biophys Res Commun 1994; 202:1426–1430.
121 de Vos S, Miller CW, Takeuchi S et al. Alterations of CDKN2 (p16) in non–small cell lung cancer. Genes Chromosom Cancer 1995; 14:164–170.
122 Kelley MJ, Nakagawa K, Steinberg SM et al. Differential inactivation of CDKN2 and Rb protein in non–small–cell and small–cell lung cancer cell lines. J Natl Cancer Inst 1995; 87:756–761.
123 Nakagawa K, Conrad NK, Williams JP et al. Mechanism of inactivation of CDKN2 and MTS2 in non–small cell lung cancer and association with advanced stage. Oncogene 1995; 11:1843–1851.
124 Okamoto A, Hussain SP, Hagiwara K et al. Mutations in the p16[INK4/MTS1/CDKN2], p15[INK4B/MTS2], and p18 genes in primary and metastatic lung cancer. Cancer Res 1995; 55:1448–1451.
125 Washimi O, Nagatake M, Osada H et al. In vivo occurrence of p16 (MTS1) and p15 (MTS2) alterations preferentially in non–small cell lung cancers. Cancer Res 1995; 55:514–517.
126 Xiao S, Li D, Corson JM et al. Codeletion of p15 and p16 genes in primary non–small cell lung carcinoma. Cancer Res 1995; 55:2968–2971.
127 Rusin MR, Okamoto A, Chorazy M et al. Intragenic mutations of the p16(INK4), p15(INK4B) and p18 genes in primary non–small–cell lung cancers. Int J Cancer 1996; 65:734–739.
128 Hussussian CJ, Struewing JP, Goldstein AM et al. Germline p16 mutations in familial melanoma. Nat Genet 1994; 8:15–21.
129 Goldstein AM, Fraser MC, Struewing JP et al. Increased risk of pancreatic cancer in melanoma–prone kindreds with p16[INK4] mutations. N Engl J Med 1995; 333:970–974.
130 Gruis NA, van der Velden PA, Sandkuijl LA et al. Homozygotes for CDKN2 (p16) germline mutation in Dutch familial melanoma kindreds. Nat Genet 1995; 10:351–353.
131 Gruis NA, Sandkuijl LA, van der Velden PA et al. CDKN2 explains part of the clinical phenotype in Dutch familial atypical multiple–mole melanoma (FAMMM) syndrome families. Melanoma Res 1995; 5:169–177.

132 Holland EA, Beaton SC, Becker TM et al. Analysis of the p16 gene, CDKN2, in 17 Australian melanoma kindreds. Oncogene 1995; 11:2289–2294.
133 Liu L, Lassam NJ, Slingerland JM et al. Germline $p16^{INK4A}$ mutation and protein dysfunction in a family with inherited melanoma. Oncogene 1995; 11:405–412.
134 Walker GJ, Hussussian CJ, Flores JF et al. Mutations of the CDKN2/$p16^{INK4}$ gene in Australian melanoma kindreds. Hum Mol Genet 1995; 4:1845–1852.
135 Borg A, Johannsson U, Johannsson O et al. Novel germline p16 mutation in familial malignant melanoma in southern Sweden. Cancer Res 1996; 56:2497–2500.
136 Gonzalez–Zulueta M, Bender CM, Yang AS et al. Methylation of the 5' CpG island of the p16/CDKN2 tumor suppressor gene in normal and transformed human tissues correlates with gene silencing. Cancer Res 1995; 55:4531–4535.
137 Herman JG, Merlo A, Mao L et al. Inactivation of the CDKN2/p16/MTS1 gene is frequently associated with aberrant DNA methylation in all common human cancers. Cancer Res 1995; 55:4525–4530.
138 Little M, Wainwright B. Methylation and p16: suppressing the suppressor. Nature Medicine 1995; 1:633–634.
139 Merlo A, Herman JG, Mao L et al. 5' CpG island methylation is associated with transcriptional silencing of the tumour suppressor p16/CDKN2/MTS1 in human cancers. Nature Medicine 1995; 1:686–692.
140 Otterson GA, Khleif SN, Chen W et al. CDKN2 gene silencing in lung cancer by DNA hypermethylation and kinetics of $p16^{INK4}$ protein induction by 5–aza 2'deoxycytidine. Oncogene 1995; 11:1211–1216.
141 Shapiro GI, Park JE, Edwards CD et al. Multiple mechanisms of $p16^{INK4A}$ inactivation in non–small cell lung cancer cell lines. Cancer Res 1995; 55:6200–6209.
142 Costello JF, Berger MS, Huang HS et al. Silencing of p16/CDKN2 expression in human gliomas by methylation and chromatin condensation. Cancer Res 1996; 56:2405–2410.
143 Lo KW, Cheung ST, Leung SF et al. Hypermethylation of the p16 gene in nasopharyngeal carcinoma. Cancer Res 1996; 56:2721–2725.
144 Glendening JM, Flores JF, Walker GJ et al. Homozygous loss of the $p15^{INK4B}$ gene (and not the $p16^{INK4}$ gene) during tumor progression in a sporadic melanoma patient. Cancer Res 1995; 55:5531–5535.
145 Gombart AF, Morosetti R, Miller CW et al. Deletions of the cyclin–dependent kinase inhibitor genes $p16^{INK4A}$ and $p15^{INK4B}$ in non–Hodgkin's lymphomas. Blood 1995; 86:1534–1539.
146 Hatta Y, Hirama T, Miller CW et al. Homozygous deletions of the p15 (MTS2) and p16 (CDKN2/MTS1) genes in adult T–cell leukemia. Blood 1995; 85:2699–2704.
147 Izumoto S, Arita N, Ohnishi T et al. Homozygous deletions of $p16^{INK4A/MTS1}$ and $p15^{INK4B/MTS2}$ genes in glioma cells and primary glioma tissues. Cancer Lett 1995; 97:241–247.
148 Jen J, Harper JW, Bigner SH et al. Deletion of p16 and p15 genes in brain tumors. Cancer Res 1994; 54:6353–6358.
149 Kawamata N, Miller CW, Koeffler HP. Molecular analysis of a family of cyclin–dependent kinase inhibitor genes (p15/MTS2/INK4b and p18/INK4c) in non–small cell lung cancers. Mol Carcinog 1995; 14:263–268.
150 Koduru PR, Zariwala M, Soni M et al. Deletion of cyclin–dependent kinase 4 inhibitor genes P15 and P16 in non–Hodgkin's lymphoma. Blood 1995; 86:2900–2905.
151 Nakamaki T, Kawamata N, Schwaller J et al. Structural integrity of the cyclin–dependent kinase inhibitor genes, p15, p16 and p18 in myeloid leukaemias. Br J Haematol 1995; 91:139–149.
152 Orlow I, Lacombe L, Hannon GJ et al. Deletion of the p16 and p15 genes in human bladder tumors. J Natl Cancer Inst 1995; 87:1524–1529.
153 Packenham JP, Taylor JA, Anna CH et al. Homozygous deletions but no sequence mutations in coding regions of p15 or p16 in human primary bladder tumors. Mol Carcinog 1995; 14:147–151.
154 Siebert R, Willers CP, Schramm A et al. Homozygous loss of the MTS1/p16 and MTS2/p15 genes in lymphoma and lymphoblastic leukaemia cell lines. Br J Haematol 1995; 91:350–354.
155 Stone S, Dayananth P, Jiang P et al. Genomic structure, expression and mutational analysis of the P15 (MTS2) gene. Oncogene 1995; 11:987–991.

156 Xiao S, Li D, Vijg J et al. Codeletion of p15 and p16 in primary malignant mesothelioma. Oncogene 1995; 11:511–515.
157 Tenan M, Benedetti S, Finocchiaro G. Deletion and transfection analysis of the p15/MTS2 gene in malignant gliomas. Biochem Biophys Res Commun 1995; 217:195–202.
158 Williamson MP, Elder PA, Shaw ME et al. p16 (CDKN2) is a major deletion target at 9p21 in bladder cancer. Hum Mol Genet 1995; 4:1569–1577.
159 Zhou X, Suzuki H, Shimada Y et al. Genomic DNA and messenger RNA expression alterations of the CDKN2B and CDKN2 genes in esophageal squamous carcinoma cell lines. Genes Chromosom Cancer 1995; 13:285–290.
160 Borgonovo Brandter L, Heyman M, Rasool O et al. $p16^{INK4}/p15^{INK4b}$ gene inactivation is a frequent event in malignant T-cell lines. Eur J Haematol 1996; 56:313–318.
161 Hangaishi A, Ogawa S, Imamura N et al. Inactivation of multiple tumor–suppressor genes involved in negative regulation of the cell cycle, MTS1/$p16^{INK4A}$/CDKN2, MTS2/$p15^{INK4B}$, p53, and Rb genes in primary lymphoid malignancies. Blood 1996; 87:4949–4958.
162 Naumann M, Savitskaia N, Eilert C et al. Frequent codeletion of p16/MTS1 and p15/MTS2 and genetic alterations in p16/MTS1 in pancreatic tumors. Gastroenterology 1996; 110:1215–1224.
163 Zariwala M, Liu E, Xiong Y. Mutational analysis of the p16 family cyclin–dependent kinase inhibitors $p15^{INK4b}$ and $p18^{INK4c}$ in tumor–derived cell lines and primary tumors. Oncogene 1996; 12:451–455.
164 Diccianni MB, Chau LS, Batova A et al. The p16 and p18 tumor suppressor genes in neuroblastoma: implications for drug resistance. Cancer Lett 1996; 104:183–192.
165 Miller CW, Aslo A, Campbell MJ et al. Alterations of the p15, p16,and p18 genes in osteosarcoma. Cancer Genet Cytogenet 1996; 86:136–142.
166 Otsuki T, Jaffe ES, Wellmann A et al. Absence of p18 mutations or deletions in lymphoid malignancies. Leukemia 1996; 10:356–360.
167 Siebert R, Willers CP, Fossa A et al. Analysis of the novel cyclin–dependent kinase 4 and 6 inhibitor gene p18 in lymphoma and leukemia cell lines. Leuk Res 1996; 20:197–200.
168 Galaktionov K, Lee AK, Eckstein J et al. CDC25 phosphatases as potential human oncogenes. Science 1995; 269:1575–1577.
169 Galaktionov K, Jessus C, Beach D. Raf1 interaction with Cdc25 phosphatase ties mitogenic signal transduction to cell cycle activation. Genes Dev 1995; 9:1046–1058.
170 Galaktionov K, Chen X, Beach D. Cdc25 cell–cycle phosphatase as a target of c-myc. Nature 1996; 382:511–517.
171 Weinberg RA. The retinoblastoma protein and cell cycle control. Cell 1995; 81:323–330.
172 Ichimura K, Schmidt EE, Goike HM et al. Human glioblastomas with no alterations of the CDKN2A ($p16^{INK4A}$, MTS1) and CDK4 genes have frequent mutations of the retinoblastoma gene. Oncogene 1996; 13:1065–1072.
173 Otterson GA, Kratzke RA, Coxon A et al. Absence of $p16^{INK4}$ protein is restricted to the subset of lung cancer lines that retains wildtype RB. Oncogene 1994; 9:3375–3378.
174 Kamb A, Gruis NA, Weaver–Feldhaus J et al. A cell cycle regulator potentially involved in genesis of many tumor types. Science 1994; 264:436–440.
175 Spruck CH, Gonzalez–Zulueta M, Shibata A et al. p16 gene in uncultured tumours. Nature 1994; 370:183–184.
176 Tam SW, Shay JW, Pagano M. Differential expression and cell cycle regulation of the cyclin–dependent kinase 4 inhibitor $p16^{Ink4}$. Cancer Res 1994; 54:5816–5820.
177 Xu L, Sgroi D, Sterner CJ et al. Mutational analysis of CDKN2 (MTS1/$p16^{ink4}$) in human breast carcinomas. Cancer Res 1994; 54:5262–5264.
178 Berns EM, Klijn JG, Smid M et al. Infrequent CDKN2 (MTS1/p16) gene alterations in human primary breast cancer. Br J Cancer 1995; 72:964–967.
179 Brenner AJ, Aldaz CM. Chromosome 9p allelic loss and p16/CDKN2 in breast cancer and evidence of p16 inactivation in immortal breast epithelial cells. Cancer Res 1995; 55:2892–2895.
180 Rush EB, Abouezzi Z, Borgen PI et al. Analysis of MTS1/CDK4 in female breast carcinomas. Cancer Lett 1995; 89:223–226.
181 Igaki H, Sasaki H, Kishi T et al. Highly frequent homozygous deletion of the p16 gene in esophageal cancer cell lines. Biochem Biophys Res Commun 1994; 203:1090–1095.

182 Liu Q, Yan YX, McClure M et al. MTS–1 (CDKN2) tumor suppressor gene deletions are a frequent event in esophagus squamous cancer and pancreatic adenocarcinoma cell lines. Oncogene 1995; 10:619–622.
183 Nobori T, Miura K, Wu DJ et al. Deletions of the cyclin–dependent kinase–4 inhibitor gene in multiple human cancers. Nature 1994; 368:753–756.
184 Gomi A, Sakai R, Ogawa S et al. Frequent loss of the cyclin–dependent kinase–4 inhibitor gene in human gliomas. Jpn J Cancer Res 1995; 86:342–346.
185 Yeudall WA, Crawford RY, Ensley JF et al. MTS1/CDK4I is altered in cell lines derived from primary and metastatic oral squamous cell carcinoma. Carcinogenesis 1994; 15:2683–2686.
186 Zhang SY, Klein–Szanto AJ, Sauter ER et al. Higher frequency of alterations in the p16/CDKN2 gene in squamous cell carcinoma cell lines than in primary tumors of the head and neck. Cancer Res 1994; 54:5050–5053.
187 Lo KW, Huang DP, Lau KM. p16 gene alterations in nasopharyngeal carcinoma. Cancer Res 1995; 55:2039–2043.
188 Lydiatt WM, Murty VV, Davidson BJ et al. Homozygous deletions and loss of expression of the CDKN2 gene occur frequently in head and neck squamous cell carcinoma cell lines but infrequently in primary tumors. Genes Chromosom Cancer 1995; 13:94–98.
189 Kinoshita H, Yamada H, Ogawa O et al. Contribution of chromosome 9p21–22 deletion to the progression of human renal cell carcinoma. Jpn J Cancer Res 1995; 86:795–799.
190 Zhou M, Gu L, James CD et al. Homozygous deletions of the CDKN2 (MTS1/p16^{ink4}) gene in cell lines established from children with acute lymphoblastic leukemia. Leukemia 1995; 9:1159–1161.
191 Okamoto A, Demetrick DJ, Spillare EA et al. Mutations and altered expression of p16^{INK4} in human cancer. Proc Natl Acad Sci USA 1994; 91:11045–11049.
192 Ohta M, Nagai H, Shimizu M et al. Rarity of somatic and germline mutations of the cyclin–dependent kinase 4 inhibitor gene, CDK4I, in melanoma. Cancer Res 1994; 54:5269–5272.
193 Luca M, Xie S, Gutman M et al. Abnormalities in the CDKN2 (p16$^{INK4/MTS-1}$) gene in human melanoma cells: relevance to tumor growth and metastasis. Oncogene 1995; 11:1399–1402.
194 Pollock PM, Yu F, Qiu L et al. Evidence for u.v. induction of CDKN2 mutations in melanoma cell lines. Oncogene 1995; 11:663–668.
195 Cheng JQ, Jhanwar SC, Klein WM et al. p16 alterations and deletion mapping of 9p21–p22 in malignant mesothelioma. Cancer Res 1994; 54:5547–5551.
196 Beltinger CP, White PS, Sulman EP et al. No CDKN2 mutations in neuroblastomas. Cancer Res 1995; 55:2053–2055.
197 Raffel C, Ueki K, Harsh GR et al. The multiple tumor suppressor 1/cyclin–dependent kinase inhibitor 2 gene in human central nervous system primitive neuroectodermal tumor. Neurosurgery 1995; 36:971–974; discussion 974–975.
198 Rodabaugh KJ, Biggs RB, Qureshi JA et al. Detailed deletion mapping of chromosome 9p and p16 gene alterations in human borderline and invasive epithelial ovarian tumors. Oncogene 1995; 11:1249–1254.
199 Caldas C, Hahn SA, da Costa LT et al. Frequent somatic mutations and homozygous deletions of the p16 (MTS1) gene in pancreatic adenocarcinoma. Nat Genet 1994; 8:27–32.
200 Chen ZH, Zhang H, Savarese TM. Gene deletion chemoselectivity: codeletion of the genes for p16(INK4), methylthioadenosine phosphorylase, and the alpha– and beta interferons in human pancreatic cell carcinoma lines and its implications for chemotherapy. Cancer Res 1996; 56:1083–1090.
201 Huang L, Goodrow TL, Zhang SY et al. Deletion and mutation analyses of the P16/MTS-1 tumor suppressor gene in human ductal pancreatic cancer reveals a higher frequency of abnormalities in tumor–derived cell lines than in primary ductal adenocarcinomas. Cancer Res 1996; 56:1137–1141.
202 Komiya A, Suzuki H, Aida S et al. Mutational analysis of CDKN2 (CDK4I/MTS1) gene in tissues and cell lines of human prostate cancer. Jpn J Cancer Res 1995; 86:622–625.
203 Ueki K, Rubio MP, Ramesh V et al. MTS1/CDKN2 gene mutations are rare in primary human astrocytomas with allelic loss of chromosome 9p. Hum Mol Genet 1994; 3:1841–1845.
204 Walker DG, Duan W, Kaye AH et al. Homozygous deletions of the MTS1 gene are rare in non–astrocytic brain tumors. Biochem Biophys Res Commun 1995; 211:404–409.

205 Ueki K, Ono Y, Henson JW et al. CDKN2/p16 or RB alterations occur in the majority of glioblastomas and are inversely correlated. Cancer Res 1996; 56:150–153.
206 Cairns P, Mao L, Merlo A et al. Rates of p16 (MTS1) mutations in primary tumors with 9p loss. Science 1994; 265:415–417.
207 Kai M, Arakawa H, Sugimoto Y et al. Infrequent somatic mutation of the MTS1 gene in primary bladder carcinomas. Jpn J Cancer Res 1995; 86:249–251.
208 Okajima E, Fukuda T, Okita S et al. Infrequent somatic alteration of p16/MTS1 in human primary superficial bladder cancers. Cancer Lett 1996; 103:227–231.
209 Quesnel B, Fenaux P, Philippe N et al. Analysis of p16 gene deletion and point mutation in breast carcinoma. Br J Cancer 1995; 72:351–353.
210 Mori T, Miura K, Aoki T et al. Frequent somatic mutation of the MTS1/CDK4I (multiple tumor suppressor/cyclin–dependent kinase 4 inhibitor) gene in esophageal squamous cell carcinoma. Cancer Res 1994; 54:3396–3397.
211 Zhou X, Tarmin L, Yin J et al. The MTS1 gene is frequently mutated in primary human esophageal tumors. Oncogene 1994; 9:3737–3741.
212 Igaki H, Sasaki H, Tachimori Y et al. Mutation frequency of the p16/CDKN2 gene in primary cancers in the upper digestive tract. Cancer Res 1995; 55:3421–3423.
213 Esteve A, Martel–Planche G, Sylla BS et al. Low frequency of p16/CDKN2 gene mutations in esophageal carcinomas. Int J Cancer 1996; 66:301–304.
214 Giani C, Finocchiaro G. Mutation rate of the CDKN2 gene in malignant gliomas. Cancer Res 1994; 54:6338–6339.
215 Dreyling MH, Bohlander SK, Adeyanju MO et al. Detection of CDKN2 deletions in tumor cell lines and primary glioma by interphase fluorescence in situ hybridization. Cancer Res 1995; 55:984–988.
216 Sato K, Schauble B, Kleihues P et al. Infrequent alterations of the p15, p16, CDK4 and cyclin D1 genes in non–astrocytic human brain tumors. Int J Cancer 1996; 66:305–308.
217 Sun Y, Hildesheim A, Lanier AE et al. No point mutation but decreased expression of the p16/MTS1 tumor suppressor gene in nasopharyngeal carcinomas. Oncogene 1995; 10:785–788.
218 Newcomb EW, Rao LS, Giknavorian SS et al. Alterations of multiple tumor suppressor genes (p53 (17p13), $p16^{INK4}$ (9p21), and DBM (13q14)) in B–cell chronic lymphocytic leukemia. Mol Carcinog 1995; 14:141–146.
219 Serra A, Gottardi E, Della Ragione F et al. Involvement of the cyclin–dependent kinase–4 inhibitor (CDKN2) gene in the pathogenesis of lymphoid blast crisis of chronic myelogenous leukaemia. Br J Haematol 1995; 91:625–629.
220 Uchida T, Watanabe T, Kinoshita T et al. Mutational analysis of the CDKN2 ($MTS1/p16^{ink4A}$) gene in primary B-cell lymphomas. Blood 1995; 86:2724–2731.
221 Shimizu T, Sekiya T. Loss of heterozygosity at 9p21 loci and mutations of the MTS1 and MTS2 genes in human lung cancers. Int J Cancer 1995; 63:616–620.
222 Ohta M, Berd D, Shimizu M et al. Deletion mapping of chromosome region 9p21–p22 surrounding the CDKN2 locus in melanoma. Int J Cancer 1996; 65:762–767.
223 Platz A, Ringborg U, Lagerlof B et al. Mutational analysis of the CDKN2 gene in metastases from patients with cutaneous malignant melanoma. Br J Cancer 1996; 73:344–348.
224 Puig S, Ruiz A, Lazaro C et al. Chromosome 9p deletions in cutaneous malignant melanoma tumors: the minimal deleted region involves markers outside the p16 (CDKN2) gene. Am J Hum Genet 1995; 57:395–402.
225 Campbell IG, Beynon G, Davis M et al. LOH and mutation analysis of CDKN2 in primary human ovarian cancers. Int J Cancer 1995; 63:222–225.
226 Hatta Y, Hirama T, Takeuchi S et al. Alterations of the p16 (MTS1) gene in testicular, ovarian, and endometrial malignancies. J Urol 1995; 154:1954–1957.
227 Bartsch D, Shevlin DW, Tung WS et al. Frequent mutations of CDKN2 in primary pancreatic adenocarcinomas. Genes Chromosom Cancer 1995; 14:189–195.
228 Sakata K, Tamura G, Maesawa C et al. Loss of heterozygosity on the short arm of chromosome 9 without p16 gene mutation in gastric carcinomas. Jpn J Cancer Res 1995; 86:333–335.

Chapter 9

Perspective

Cyclin–CDK Regulation is Cell Cycle Regulation

Undoubtedly one of the most significant achievements in the biological sciences in recent times has been the monumental leap forward in our understanding of the cell cycle. Following a number of convergent genetic and biochemical lines of evidence, we have uncovered the keystone of the regulatory structure through which cell cycle events are mediated—the cyclin/cyclin–dependent kinase (CDK) complex. This unique species of kinase controls the checkpoints that the cell must satisfy before progressing through the phase transitions of the cell cycle. Any stimulus that brings about a change in a cell's proliferative status must through some means impinge upon a cyclin–CDK complex. For it is clear that this complex is the primary "switch" to which the cell is tuned; an active complex is "on," pushing the cell forward by phosphorylating a host of substrates (many of which are as of yet unknown), and an inactive complex is "off," arresting the cell's progress through the cell cycle. Upon analysis of the ample (and growing) body of evidence concerned with the cell cycle, we conclude that there are (at least) four mechanisms through which the cyclin–CDK "switch" may be alternately shifted between the "on" and "off" states. Let us briefly summarize these mechanisms.

1. The assembly/disassembly of the cyclin–CDK complex.

Prerequisite for the kinase activity of the complex is the presence of both the cyclin and CDK subunits. CDKs are present in more or less consistent levels throughout the cell cycle, but the defining characteristic of cyclins is that they are periodic and phase–specific, i.e., their levels oscillate during the course of the cell cycle and peak during particular intervals of the cycle. For instance, the cyclin B level is lowest at the start of G1 phase, increases during the S and G2 phases, and reaches a maximum at the start of M phase. By its very nature the cyclin exerts a constraint on cyclin–CDK activity, for only when the cyclin level is at its highest are there enough cyclin subunits to stoichiometrically associate with the CDK subunits and give rise to cyclin–CDK complexes. Inversely, when the cyclin level drops, the cyclin–CDK complexes can no longer be maintained. The cyclin B level precipitously drops during M phase through ubiquitin–mediated degradation of the cyclin, with the consequent disassociation and inactivation of cyclin B–CDK1 complexes. The accumulation and destruction of cyclin B thus defines a window during which cyclin B–CDK1 complexes can appear. It is only during G2 phase and early M phase that we should have any expectation of observing cyclin B–CDK1 activity. Moreover, the onset of mitosis must take place during this interval, since cyclin B-CDK1, also known as maturation–promoting factor (MPF), is the primary mitotic inducer.

This mechanism therefore provides two handles through which cyclin–CDK activity may be manipulated. First, varying the rate of synthesis of the cyclin protein alternatively accelerates or delays the timing of cyclin–CDK activity. Similarly, boosting or inhibiting the processes responsible for the degradation of the cyclin subunit modulates the inactivation of the complex.

2. The change in the phosphorylation state of one or two residues near the N-terminus of the CDK.

In yeast the crucial residue appears to be a single tyrosine; in vertebrates, a pair of adjacent threonine and tyrosine residues. Even if the cyclin subunit is present in the cyclin–CDK complex, the complex will not be active unless these residues are unphosphorylated. To give an illustrative example, CDK1 is phosphorylated on these residues during much of the cell cycle. Although a significant number of cyclin B–CDK1 complexes are formed during G2 phase, they remain inactive (the so–called "pre–MPF" complexes) until the CDK1 subunits are dephosphorylated at the end of G2 phase. Having cyclin–CDK activity dependent on the phosphorylation state of the N-terminal residues provides another constraint that more tightly defines the interval during which the cyclin–CDK "switch" is allowed to be turned "on."

Whether the residues are phosphorylated or unphosphorylated evidently depends on the status of a tug–of–war between protein kinases and phosphatases. The Wee1 kinase (and its relatives) directly phosphorylate the residues, preventing the activation of the cyclin–CDK complex. The CDC25 phosphatase acts as an antagonist, directly dephosphorylating the residues and promoting cyclin–CDK activity. Both proteins are subject to activation/inactivation by other kinases and phosphatases, which presumably are themselves subject to regulation. Thus there exists a network of kinases and phosphatases convergent upon the CDK. Cyclin–CDK activity can be manipulated in an indirect fashion by reinforcing or abrogating the activity of one or more of the components of the upstream network, swaying the balance in favor of either Wee1 (CDK phosphorylation) or CDC25 (CDK dephosphorylation).

3. The change in the phosphorylation state of a residue in the T-loop of the CDK.

The crucial residue is a specific threonine conserved in all well–studied organisms, from yeasts to humans. In contrast to the N-terminal residues, this threonine must be phosphorylated in order for the cyclin–CDK complex to have kinase activity. Returning to our example, the appropriate threonine residue in CDK1 is unphosphorylated in G1 phase. By the time the cell has progressed into late G2 phase, the residue has been phosphorylated, allowing cyclin B-CDK1 to assume kinase activity and induce mitosis. The T–loop threonine residue thereby represents yet another constraint on the timing of CDK activation.

It is presumed that the phosphorylation state of this crucial threonine residue is regulated by an antagonistic kinase/phosphatase pair (to date the phosphatase has not been identified). Remarkably, the kinase is none other than a cyclin–CDK complex (cyclin H–CDK7). The presence of this cyclin–CDK cascade offers a richly expanded realm of possibilities in regards to cyclin–CDK regulation, for we may suppose that cyclin H–CDK7 is itself regulated by each of the four mechanisms we are presently discussing. If we posit the existence of a network of kinases and phosphatases converging upon cyclin H–CDK7 and, in turn, our prototype cyclin–CDK complex, we see that there are a large number of avenues through which the status of the cyclin–CDK complex may be manipulated.

4. The physical association of inhibitor molecules or enhancer molecules with the cyclin–CDK complex.

The growing family of regulators known as CDK inhibitors (CKIs) seem to operate by binding the complex and, in doing so, preventing the CDK subunit from acting upon substrates, or by even breaking up the complex altogether. Even if the other three prerequisite conditions for cyclin–CDK activity were satisfied—the presence of the cyclin subunit, the dephosphorylation of the N–terminal residues, and the phosphorylation of the T-loop residue—the presence of the CKI would be sufficient to nullify the kinase activity. CKIs signify a fourth constraint on cyclin–CDK activity. Any environmental signal that stimulated the production of the appropriate CKI would block cyclin–CDK function; conversely, if another stimulus repressed the CKI, the cyclin–CDK complex would be freed from that restraint.

It has recently emerged that there are additional proteins that bind to the cyclin–CDK complex; their functions are unclear, but they do not seem to act as inhibitors. The possibility must be held out that they serve to enhance cyclin–CDK activity. One can imagine the ways in which such enhancers might operate: by gluing together the cyclin and CDK subunits, by conformationally facilitating the binding of substrates, or even by titrating out CKIs and blocking their inhibitory effects. Enhancers would provide yet another handle through which cyclin–CDK complexes would be subject to manipulation.

Each of the four mechanisms we have just described individually gives a simple prescription for how the cell cycle might be controlled. Together, they constitute an interwoven regulatory fabric of surprising complexity. For a typical protein kinase there is just one regulatory mechanism, phosphorylation/dephosphorylation by upstream kinases and phosphatases; the fact that there are at least four distinct ways to regulate a cyclin–CDK complex is confirmation of the crucial role it plays in the cellular milieu. We are justified in saying that with few exceptions every agonist, stimulus, or cellular signal that bears upon cell proliferation acts in some way through a cyclin–CDK complex.

The Current Model of the Cell Cycle

Our present–day understanding of the cell cycle relies heavily on cyclin–CDK complexes. Let us take a brief tour (constructed from evidence amassed by investigators over the past few years and by no means close to a complete formulation) from early G1 phase to late M phase in a typical, normally dividing mammalian cell (Fig. 9.1).

At the start of G1 phase we witness a newly formed, growing cell. For the most part cyclin levels are low, and though CDKs are present no cyclin–CDK complexes can be formed. As G1 progresses the D–type cyclins rise in quantity and bind preferentially to CDK4 and CDK6. These complexes remain inactive until conditions in the cell are appropriate for DNA synthesis. The cell has reached what is called the restriction (R) point (in yeast it is known as START), the checkpoint which, once passed, commits a cell to duplicating its chromosomes and carrying out the subsequent mitosis. The cyclin D–CDK4 and –CDK6 complexes are eventually dephosphorylated by a CDC25 family member and phosphorylated by cyclin H–CDK7; these actions signify that the cell is surpassing the R point. The now–active complexes phosphorylate the retinoblastoma susceptibility gene product (pRb), inducing a conformational change that prevents the latter from binding to the transcription factor E2F. Freed from pRb's restraint, E2F induces the production of a number of proteins that prepares the cell for DNA synthesis, including cyclin E. As the cyclin E level peaks, active cyclin E-CDK2

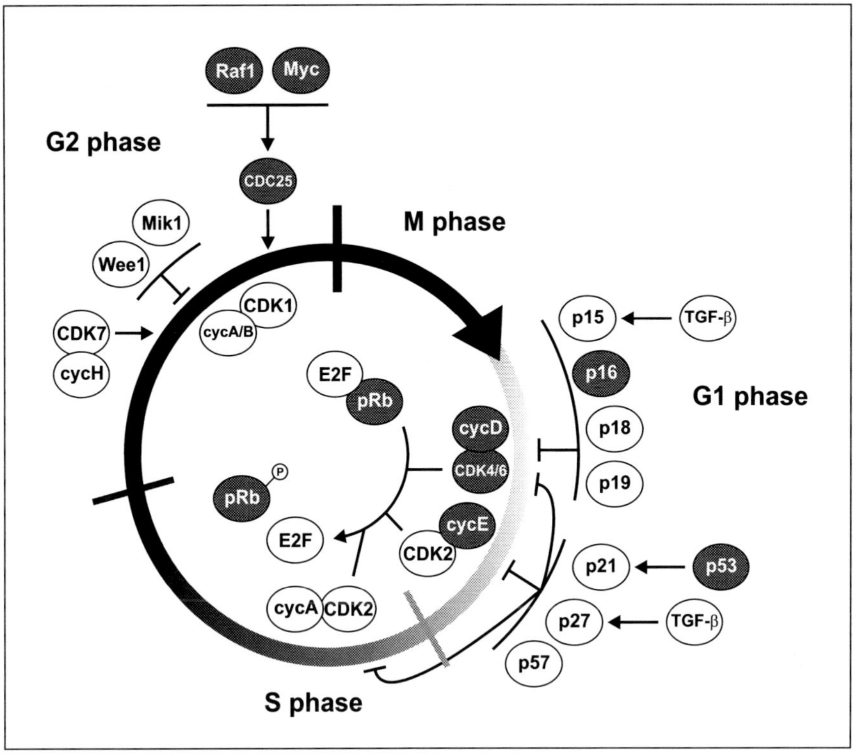

Fig. 9.1. Diagram depicting a generic cell cycle as regulated by cyclins, CDKs, CKIs and other control proteins. Arrows indicate an activation or upregulation of the target protein, and bars indicate an inhibition or downregulation of the target protein. The proteins depicted as grey ovals, in contrast to those depicted as white ovals, have been found to be mutated or deleted in a number of tumor types. This figure is a synthesis of information obtained from a limited number of biological systems and should not be considered to universally apply to all species. For simplicity's sake many details have regrettably been omitted; for instance, while cyclin H–CDK7, Wee1/Mik1, and CDC25 are shown acting exclusively upon cyclin A- and cyclin B–CDK1 during G2 phase, they are thought to act upon cyclin D– and cyclin E–CDK complexes during G1 phase with equal efficacy.

complexes are generated, which in turn hyperphosphorylate pRb, reinforce the effects of E2F, and presumably phosphorylate a number of downstream targets that allow DNA replication to commence. The cell proceeds into S phase.

By this time the cyclin A level has increased to the extent that active cyclin A–CDK2 complexes appear. These kinase complexes maintain the phosphorylation of pRb (by this time both cyclins D and E have been degraded) and presumably sustain DNA replication. As S phase segues into G2 phase, cyclin A-CDK1 complexes assemble while cyclin A–CDK2 remains active, each phosphorylating substrates in preparation for mitosis. It is cyclin B, however, that is the primary mitotic inducer. By the end of G2 phase the cyclin B level has peaked and large quantities of inactive cyclin B–CDK1, or "pre–MPF," have been generated. During G2 phase cyclin H–CDK7 phosphorylates the T-loop threonine residue of CDK1; nevertheless the onset of mitosis cannot occur until CDC25 dephosphorylates the N–terminal residues at a faster rate than Wee1 phosphorylates them. The delicate balance between the kinase and phosphatase is swayed in favor of CDC25 only when all the necessary conditions are met, i.e., the chromosomes have been fully duplicated, there are enough nutrients to carry out cell division, etc. Active cyclin B–CDK1 phosphorylates and activates CDC25, giving rise to a positive feedback loop that converts the entire pool of pre–MPF into active MPF. Cyclin B–CDK1 jumpstarts M phase by phosphorylating a variety of targets important in the mitotic process.

M phase witnesses the calculated destruction of cyclins A and B via ubiquitin–mediated proteolysis. As the levels of these cyclins are depleted, MPF unravels and the impetus for cell division is removed. By this time the original cell has successfully divided into two progeny cells, and the cycle begins anew.

At any time during this process the cell can be brought to a halt if CKIs conspire to block the ability of the active cyclin–CDK complex to phosphorylate its substrates. For example, DNA damage caused by ultraviolet radiation exposure during G1 phase induces the rapid expression of the p53 tumor suppressor protein. p53 in turn motivates the production of the CKI p21, which binds and inactivates cyclin D-CDK4 and –CDK6 complexes. The cell is thus arrested at the R point, allowing time for the DNA damage to be repaired before the cell commits to a course of reproduction.

As discussed before, the four mechanisms for the regulation of cyclin–CDKs represent four constraints on the cell as it progresses through the cell cycle. Indeed, it might be said that they *define* the cell cycle. The requirement that cyclin be bound to a CDK for the latter to have kinase activity establishes the order of events during the cell cycle; mitosis must happen after the initiation of DNA synthesis because the cyclin E level peaks before the cyclin B level. The dual requirements of dephosphorylation of the CDK's two crucial N-terminal residues and phosphorylation of the crucial T-loop threonine residue determine the precise, appropriate timing of events such as the onset of mitosis. And the ability of CKIs to preemptively block cyclin–CDK activity allows the cell to immediately arrest in response to such threats as DNA damage. The cyclin–CDK paradigm is clearly the bounty of eons of evolution that renders the cell extraordinarily responsive to its environment.

Targets for Clinical Intervention in the Treatment of Cancer

There are nevertheless times when the cell acts without regard to its environment. The result is inappropriate cell proliferation, a condition that all too easily can give rise to cancer. Since cyclin–CDK regulation is central to the progression of the cell cycle, it should not be surprising that mutations in cyclins, CDKs and their regulators are implicated in carcinogenesis.

Cancer is unfortunately an immensely complex disease that defies any simple prescription for treatment. There are two complications in particular that have made developing a comprehensive therapy an extreme challenge, with the consequence that cancer has been an almost insurmountable foe for modern medicine. First, cancer arises when enough cell cycle regulators have suffered mutations. However, there is no universal set of mutations which are required for carcinogenesis to take place. It is entirely possible for one cancer to have owed its existence to a set of five mutated regulators, and for another cancer to have developed from a mutually exclusive set of five mutations. There is one general feature shared among all cancers—that the cell cycle has been deregulated—but there are no specific features characteristic of all cancers. There is no one mutation that we can exploit in developing a cure for cancer. In this sense, cancer is not a singular disease but rather a panoply of diseases, each of which demands a custom–tailored cure.

The second complication is that cancerous cells are not so much "diseased" cells as they are "misbehaved" cells. The cell cycle machinery in cancerous cells is no different than that of normal cells; it is simply acting inappropriately, locked in continuous "play." Any treatment that targets some component of the machinery in cancerous cells will therefore run the risk of unleashing unwanted side effects on normally dividing cells. Indeed, present–day chemotherapies are often as harmful to the patient as the cancer itself. The ideal therapy must achieve specificity, pinpointing and destroying cancerous cells while leaving the rest of the body's cells untouched.

Having noted these obstacles in devising treatments for cancer, let us now frame the question of how the four mechanisms of cyclin–CDK regulation, when deregulated, can contribute to carcinogenesis. We will focus on the specific regulators whose mutations have been found to have a high correlation with cancer.

1. The assembly/disassembly of the cyclin–CDK complex.

The amplification of D-type cyclins, particularly cyclin D1, is a hallmark of a number of cancers. An overabundance of cyclin D during G1 phase allows for the untimely assembly of cyclin D-CDK4 and –CDK6 complexes. Assuming that other conditions were appropriate (dephosphorylation of N-terminal residues, absence of CKIs, etc.), these active complexes would drive the cell through the

R point prematurely. The cell would thus be committed to a round of DNA replication and mitosis earlier than a normally growing cell would be.

2. The change in the phosphorylation state of adjacent threonine and tyrosine residues near the N-terminus of the CDK.

There is evidence that CDC25, the phosphatase responsible for dephosphorylating the N-terminal residues of the CDK, is the product of a proto–oncogene. It is conceivable that constitutively active CDC25 could drive the cell through various checkpoints in an untimely fashion. For example, any complexes that were generated in G1 phase by the association of cyclin E and CDK2 would be immediately activated by CDC25 (assuming concurrent phosphorylation by cyclin H-CDK7 and no interference from CKIs). The premature presence of active cyclin E-CDK2 would instigate the too early onset of S phase.

3. The change in the phosphorylation state of the T-loop threonine residue of the CDK.

There is no evidence that cyclin H-CDK7 has any role in cancer, and not enough is known about the complementary phosphatase to implicate it in oncogenesis.

4. The physical association of inhibitor molecules with the cyclin–CDK complex.

The CKI p16 is deleted in a number of tumors, eliminating much of the cell's ability to respond to negative growth signals. If a cell lacking functional p16 has generated active cyclin–CDK complexes (and no other CKIs are being expressed in sufficient quantity), the cell will be unable to prevent itself from passing through checkpoints. A similar situation arises when CDK4 is mutated such that it cannot bind to p16, a condition found in a small but significant number of cancer cases. p16 will be unable to restrain the cell from progressing through the R point and committing to an eventual mitosis.

Cyclin D, CDK4, p16, and CDC25 (and possibly cyclin E, CDK6, and p15) are the cell cycle regulators apparently involved in cancer formation. All of these proteins have been shown to function upstream of pRb, which itself is commonly mutated in tumors. Together these molecules constitute a pathway suspected to be altered in the majority (if not all) cases of cancer. Mutating just one of these proteins is not sufficient to convert a cell into a full–blown tumor; nevertheless, a cell carrying a mutation would more easily and quickly pass through the R point and the G1/S phase transition, and the successive cell cycles of its descendant cells would be marked by abbreviated G1 phases. Such deregulated cells would be prime candidates to be foci for nascent tumors, especially if they suffer further mutations of various oncogenes and tumor suppressor genes.

We know which cell cycle regulators are likely to play a causative role in cancer, but can we use this knowledge to devise specific cures? It will depend on our ability to target cancerous cells in the body. If a patient had a tumor of cells with deleted *p16*, we could in theory halt the growth of the tumor by delivering the *p16* gene into the cells. There are two difficulties with this plan. First, modern biology has not yet devised a functional delivery vehicle of any sort. Inventing such vehicles is the dream of those pursuing the holy grail of "gene therapy;" various viral and immunological approaches currently in trials have shown some promise, but unqualified success in this field will come years in the future. Moreover, even if we possessed an appropriate vehicle, how could we ensure that p16 was delivered solely to cancer cells? There would likely be undesirable side effects if the growth–inhibiting p16 were inserted into healthy, normally dividing cells. Such complications make gene therapy a purely theoretical solution to cancer at the present time.

The only practical solution available is that which modern medicine relies upon today: the use of exogenously delivered pharmaceuticals. If we could develop a drug that could be injected into the bloodstream and permeate into cells everywhere in the body, *and* could mimic the inhibitory effect of p16 by binding to cyclin–CDK complexes in cancer cells, *but* which only weakly interacted with complexes, such that it would act as an inhibitor in *p16*–deficient cells but would be outcompeted by p16 in healthy cells, then we would have a viable treatment (*not* a cure)—*but* only for tumors with *p16* deletion, a fraction of the totality of cancers. Development of such a drug is a long shot, and at best it would serve as just another weapon in our arsenal of chemotherapies.

It is not our intent to offer a pessimistic view of our chances in successfully waging war against cancer. On the contrary, we should be inspired by our tremendous progress in the last quarter century and our ability now to plan a rational campaign to roll back the scourge of humankind. We are no longer blindly searching for some miraculous panacea; we have reached an understanding of the molecular basis of cancer, we are developing strategies to systematically treat cancer on a case–by–case basis, and we are slowly but surely assembling the tools to employ our strategies. In another quarter century, it is conceivable that cancer will become just another dangerous predator caged by modern medicine, vanquished by human intellect.

Subject Index